Telecom Extreme Transformation

The Road to a Digital Service Provider

Kaveh Hushyar
Former Chief Engineering Officer, AT&T, Inc.
Stanford University, California, USA

Harald Braun
Former President & CEO, Aviat Networks, Inc.
President and CEO, Siemens Networks Llc., USA

Dr. Hossein Eslambolchi
Former President, CIO, CTO, AT&T Inc.

CRC Press is an imprint of the
Taylor & Francis Group, an **informa** business
A SCIENCE PUBLISHERS BOOK

First edition published 2021
by CRC Press
6000 Broken Sound Parkway NW, Suite 300, Boca Raton, FL 33487-2742

and by CRC Press
2 Park Square, Milton Park, Abingdon, Oxon, OX14 4RN

CRC Press is an imprint of Taylor & Francis Group, LLC

Library of Congress Cataloging-in-Publication Data

Names: Hushyar, Kaveh, 1952- author. | Braun, Harald (Telecommunications engineer), author. | Eslambolchi, Hossein, 1957- author.
Title: Telecom extreme transformation : the road to a digital service provider / Kaveh Hushyar, former chief engineering officer, AT&T, Inc., Stanford University, California, USA, Harald Braun, former president & CEO, Aviat Networks, Inc., president and CEO, Siemens Networks Llc., Hossein Eslambolchi, former president CIO, CTO, AT&T, Inc.
Description: First edition. | Boca Raton : CRC Press, 2021. | "A science publishers book." | Includes bibliographical references and index. | Summary: "The next wave of telecommunication and service provider transformations will be very different, we call it extreme transformation, in that the communication service providers (CSP) have to become a Digital Service Provider (DSP) to stay relevant. In the DSP world the customers are not just "human beings", but it also includes billions of sensors and IoT devices that will be revolutionizing digital lifestyle with relevant content enabled by data mining, and leading to decision making, and entertainment. for CSPs, to stay relevant, they must compete, not only with differentiation in network reliability for quad-play connectivity services, but also must move up the value chain, and to transform to become a Customer-Centric DSP for digital services. The extreme transformation from a CSP to a DSP status is what we are covering in this book"-- Provided by publisher.
Identifiers: LCCN 2021000985 | ISBN 9780367750138 (hbk)
Subjects: LCSH: Wireless communication systems--Technological innovations. | Internet of things--Forecasting. | Application service providers. | Internet service providers. | Organizational change.
Classification: LCC TK5103.2 .H865 2021 | DDC 384.3/3--dc23
LC record available at https://lccn.loc.gov/2021000985

ISBN: 978-0-367-75013-8 (hbk)

ISBN: 978-0-367-75017-6 (pbk)

ISBN: 978-1-003-16074-8 (ebk)

Typeset in Times New Roman
by Radiant Productions

Preface

Objectives

It is the ambitious purpose of this book to provide a playbook for the extreme transformation of a Communication Service Provider (CSP) to include the Digital Service Provider (DSP) status. A CSP that does not transform to a DSP, will face an "innovator's Dilemma" (Christensen, 2016) as it has created many casualties in different industries.

This book lays out the authors' experience and the learning from the transformation of many global networks of the tier 1 CSPs over time. It provides a playbook for evaluation of the legacy networks, and the transformation of its shortcomings into the Best-In-Industry (BII) global DSP with competitive performance. In this book, we address the questions about the "Why" of transformation, and the "What" of transformation. The answers to the "how-to" of transformation are not the subject of this book. The "How-to" of transformation is highly dependent on the state of the CSPs before the start of a transformation. The "how-to" is best learned and practiced on the job, motivated by the need for the transformation, and leveraging the partnership with experts with global transformation experience.

The motivation for this book is to pass on the authors' lifetime learnings and the experience for the transformation of customer experience, and the DSP platform build-out at a global scale which is sustainable and is built to last. We think it has the potential to provide a stepping stone for a much better place, from where we picked it up just a few years ago.

Acknowledgement

The idea of the book began many years ago between the three of us while working on several projects, in different companies (Service Provider, Product Manufacture), different organizational positions and definitely with different views of how things should be done, but always with one goal in mind: Get it done right, on time, within budget and with no compromises on quality! This book is a work product of this collaboration between the authors over the last 2 years while we focused on the learnings of our combined knowledge, failures, and experiences over all the projects we executed in the telecommunication industry on a global level. If this book contributes in any way to help plan and execute a transition from a CSP to a DSP, we will consider it a success.

You can't write a book like this without mentioning the mentors, tutors, and teachers who guided our professional career. Also, we got many powerful inputs from thought leaders in the industry while we were debating details in certain sections of this book. In summary, it would not have been possible to write this book without the meaningful input and commitment made by close friends, family, and colleagues.

For the Mentors and Tutors, the list has to start with Frank Ianna, former Chief Quality Officer of AT&T who is a global lead in network reliability; George Nolen, former President and CEO of Siemens North America who taught us to look at things always from different angles but foremost from the benefit we create for the end-customer, that if what we do, creates value for them, and if this can't be answered with a convincing YES, then it must create value for anybody in the value chain.

Our main teacher, when it comes to flawless execution of network operations, is John McCanuel, former AT&T Senior Vice President; when it comes to looking at subjects from a geo-economic and geostrategic perspective, our teachers include Anton Schaaf, former Group Board member of Siemens Networks and CEO Coach for southeast Asia, Joe Cook, former Senior Vice President of Verizon for Global Network Planning and Engineering, and Alberto Gross, President and CEO of CLA direct, an innovative telecommunication Solution Provider for Latin America and Caribbean markets; and last but not least, Karel Pienaar, Navi Naidoo, Hassan Motalebpoor, Edris Saberi, and Pete DaSilva were responsible for influencing our thinking for executing projects in Europe-Middle East and Asia. Our appreciation goes to Tom Rutledge, Chairman and CEO of Charter Communications. Tom's visionary view about the US telecommunication industry had a big influence on all of us. Many thanks to all for coaching us on the cultural differences between our heritage and the way topics can be approached with different cultural values and in a diverse cultural environment.

Among the many people who very early in our professions introduced us to the CLOUD based delivery concept of functionalities, we have to thank our friend and great Entrepreneur, Michael Tessler, former President and CEO of Broadsoft, who founded the company and build it to the global market leader in cloud-based unified communications solutions. Also, very influential in our way to position business solution into a global customer base were Harald Link, a Thai-German businessman, industrialist, philanthropist, and owner of B. Grimm of Bangkok, Thailand, and Tzvika Friedman, former CEO of Alvarion and currently an active investor in global innovative telecommunication business solutions. Many intensive breakfast meetings influenced our long-lasting thinking.

On the complex issue of optical Networking, we are grateful that we had very good counselors and coaching in Simon Zelinger, former Vice President of AT&T, and Dr. Hans-Juergen Schmidtke, a Communications Network Specialist experienced in optical networking solutions, now working with Facebook on their influential Telecom Infrastructure Project (TIP).

In addition, our appreciation goes to the countless hours we spend during our professional careers on multifaceted technology challenges with Behrokh Samadi, Jim Medica, Robert Covel, Chris Dressler, Nikos Theodosopoulos, and Mike Hayashi.

Contents

CHAPTER 1
Introduction

1.1 About The Authors

We the authors are the students of the telecommunication industry who have been leaders and practitioners in the space of telecom and digital services transformation. The authors in alphabetical order are:

(1) Harald Braun: former President and CEO of Aviat Networks, Inc., and the President and CEO of Siemens Networks LLC, USA. Engineering degree in Telecommunications from the University in Aachen/Germany. Prominent standards influencer and spokesman in the telecommunications industry and recognized as one of America's "Top 100 Voices of the IP Communications Industry" by Internet Telephony magazine. Serial High-Tech Entrepreneur in the Digital Economy.

(2) Hossein Eslambolchi: Former Chief Transformation Officer and president, CIO, CTO of AT&T, with more than 1400 registered patents, 2017 Ellis Island Medal of honor, Thomas Alva Edison Prestigious NJ inventor Hall of Fame, Info world Global Top 25 CTO award in 2005: PhD in Electrical Engineering from San Diego State University.

(3) Kaveh Hushyar: Former Chief Engineering Officer of AT&T. Cross industry leadership in Transformation. MS Industrial Engineering from Stanford University.

The authors are actively engaged in providing consulting services on a variety of transformational projects globally.

1.2 Who Will Benefit From This Book?

The authors have written this book for a variety of readers:

(a) Telecom and digital services professionals, from the working level all the way up to the executives, who are planning to embark on massive transformational projects, in an intensely competitive market, from a Communication Service Provider (CSP) status to include Digital Service Provider (DSP) status, or from a DSP to include CSP status; or to transform their legacy CSPs, in a non-competitive market, for efficiency and effectiveness.

(b) Telecom and technology equipment manufacturers who develop future hardware/software products to enable the transformation from a CSP to include DSP status, and vice versa.

(c) Institutional investors who are trying to evaluate and/or to establish their investment decisions in the leading-edge companies across different industries of Telecom, Equipment Manufacturers, and the Digital Service Providers.

(d) Management Consultants in Telecom and digital services who are trying to leverage a solid benchmark for transformation engagement with their clients.

(e) University students globally, majoring in telecommunication and technology products and services,who need to develop a breath of understanding for the end-to-end DSP, as well as a guide toward identifying areas for development of specific depth for their professional interest and career planning.

Given the above classes of readers for this book, one would benefit the most if equipped with the basic understanding of the telecommunication networks/services and its practices.

1.3 How This Book Is Organized

The scope of CSPs' and DSPs' business is broad, and its transformation is complex and extensive. We are planning to present the subject of Telecom Extreme Transformation in two volumes:

Volume I, the subject of this release, covers the playbook for the assessment of the legacy CSP networks, essential Best-In-Industry practices for transformation, End-to-end future and transitional architecture for a global Digital Service Provider (DSP) with the focus on unification of wireless/wireline/data networks, and the unified Services over IP (SoIP) platform; and

Volume II, the subject of future release, will cover Wireless last mile access, Security, and Operations and Business Support Systems (OSSs/BSSs).

Volume I is organized into the following chapters: Preface, including the acknowledgement; Chapter 1 (Introduction) covers the background of the authors, the potential audience for the book. It also provides the Future of Telecom (technology direction over the next 10+ years for Wireless, Wireline, Devices, and Applications/content), and a brief executive summary for the road ahead for transformation; Chapter 2 addresses the Preparation for transformation (why to transform, transformation targets, best-in-industry practices); Chapter 3 presents an Assessment of present mode of operations (PMO) in telecom (performance, Capex, Opex, business issues); Chapter 4 covers the Future Mode of Operations (FMO) in telecom (transformational targets, network and systems architecture for wireless, wireline, video, and content delivery); Chapter 5 presents the FMO integrated and real time dashboard for flawless operations of the overall network; Chapter 6 presents FMO

Network Operations and services; Chapter 7 provides a brief view of the Regulatory issues and challenges; Chapter 8 covers the Transition Mode of Operation to navigate from PMO to FMO; Chapter 9 outlines the Conclusion; Chapter 10 is an appendix for the book; and Chapter 11 outlines the references and the citation.

1.4 Future of Telecom

We believe that, by 2030, most of the traditional CSPs will be transformed to become Digital Service Providers (DSPs). Today's CSPs are now the providers of connectivity services (traditional connectivity pipes with low margin), plagued by capital intensive platforms, and slow response to meet new customer demands. These CSPs will be transformed into DSPs with highly reliable high-speed connectivity services, and vertically integrated services with high margin Digital Services, highly CAPEX/OPEX unit cost-efficient, customer-focused with BII QoE performance in a broad sense of connectivity, content, and touchpoints with the end customers and suppliers/partners.

Tomorrow's DSPs will monetize the bit stream on their connectivity network, and have the potential to take on digital disruptors, such as Google, FB, and Netflix, with strategic advantage for high-quality ETE networks. We believe that tomorrow's DSPs will become solution providers for new digital services in an age of exponential growth and digital lifestyle, as well as a role model for digital touchpoints and the capacity to connect the unconnected 3+billion people by 2030. Tomorrow's DSPs will measure their business performance by revenue per bit of traffic on their network.

Digital lifestyle and the growth of 'digital Nomads' is transforming the way we live our lives. It is empowering us to communicate anytime, anyplace, with any device on any network. It is improving our quality of life through digital touchpoints, and is providing businesses with an immense opportunity for value creation through the introduction of innovative service solutions. Digital lifestyle is enabling a new way for capital flow through the global economy.

In the new economy, "Over-The-Top"(OTT) (Journals.elsevier.com/telecommunications-policy, 2020) (Introduction to eTOM White Paper, 2017) digital services, interactive video, and user-generated content are becoming more prevalent. Players such as Facebook, Google, Apple, YouTube, Instagram, Netflix, or their future incarnations are reaping the rewards of new revenue streams being generated by advertising, payed-for content, on-demand services and application services. The emergence of the Internet-of-Things (IoT), Virtual Reality (VR), Augmented Reality (AR) and the rapid growth of autonomous vehicles will change future telecommunication networks forever.

We believe that the digital lifestyle is the force that will shape the future of telecom. The future of telecom must be seen in the cross-section of four essential elements: *Devices, Content/Applications, wireless*, and *wireline networks* as depicted in the picture below (Figure 1.1). The key question for CSPs is the ownership of this cross-section. On one extreme, CSPs will remain to be the owner of the traditional pipes with shrinking and low margin products, and on the other extreme, they become DSPs with high margin smart pipes, while leveraging the strategic asset of their network for high availability services, by offering content and/or devices. It is also true that today's content providers (such as Google) would evolve to become a DSP.

This cross-section is evolving, and based on our collective insight, by 2030, the cross-section will bring about the digital lifestyle leading to ambient solutions for the customers.

Ambient solution for consumers is where digital services are utilizing the IoT sensors' sensitivity and responsiveness to the presence of people; and for the business, it is about collaboration, big data, and sound decision making, and the availability to work from anywhere at any time.

As you see in Figure 1.1, by 2030, *Devices* will go to common robotics, fueled by IoT sensors and Robotic vision; for *mobility*, it is the scalable 7G (10 Gbps) with 100-to-1 wireless vs Wireline endpoints; and for *wireline*, it is the scalable Exabyte backbone and Yottabyte data centers fueled by the unification of the wireless and wireline networks, 100 Gbps last-mile, secure cloaked network, and on-demand reconfigurable networks, where the network is being autonomously configured in real-time to deal with catastrophic events; The *content & applications* will take us to tele-immersion and ambient intelligence solutions fueled by Cognitive Subject Matter Experts (SMEs).

At this cross-section, the people are in the center stage, for quality of life entertainment, working productively, and having fun at it too. The devices enable big data and intelligence gathering; the wireline enables the exchange of big data (east-west pipes in addition to the north-south) over the geographical environment; wireless enables extreme mobility and anyplace communication at High-speed; the applications and content provide seamless touchpoints between people and businesses, and insightful context to the big data which will facilitate human decision making as well as entertainment.

Today, the ownership of the cross-section is spread across 4 key players: (1&2) the traditional CSPs (for capital intensive & low margin pipes in *wireline and wireless networks*), (3) new and emerging companies such as Google and FB (for high margin *content, applications, and advertisement*), and (4) the traditional Network Equipment Providers, such as Apple and Cisco (for innovative and high margin *devices and network gears*).The fundamental question for the legacy CSPs is the scope of the ownership of this cross-section going forward. At the end of the day, the CSPs have to find their way into the high margin content and applications to stay relevant. Meanwhile, today's content providers will be vying for connectivity services, demanding a massive transformation to become a DSP.

There are some CSPs who will be leading the revolution, and others who will take advantage of the learning from the leaders, to build and to deliver the BII services for their customers. The leaders of the revolution will pay a significant price to be the lead,but they will become global DSPs and will be the first to benefit from the leading-edge technology and stay ahead of the competitive curve.

To deliver on this future with a competitive posture,the big league CSP players must leverage a sound set of processes and BII practices that are timeless to piece together the off-the-shelf technology components of the time, to meet and to exceed customers' expectations, and to maintain that network performance at all times.

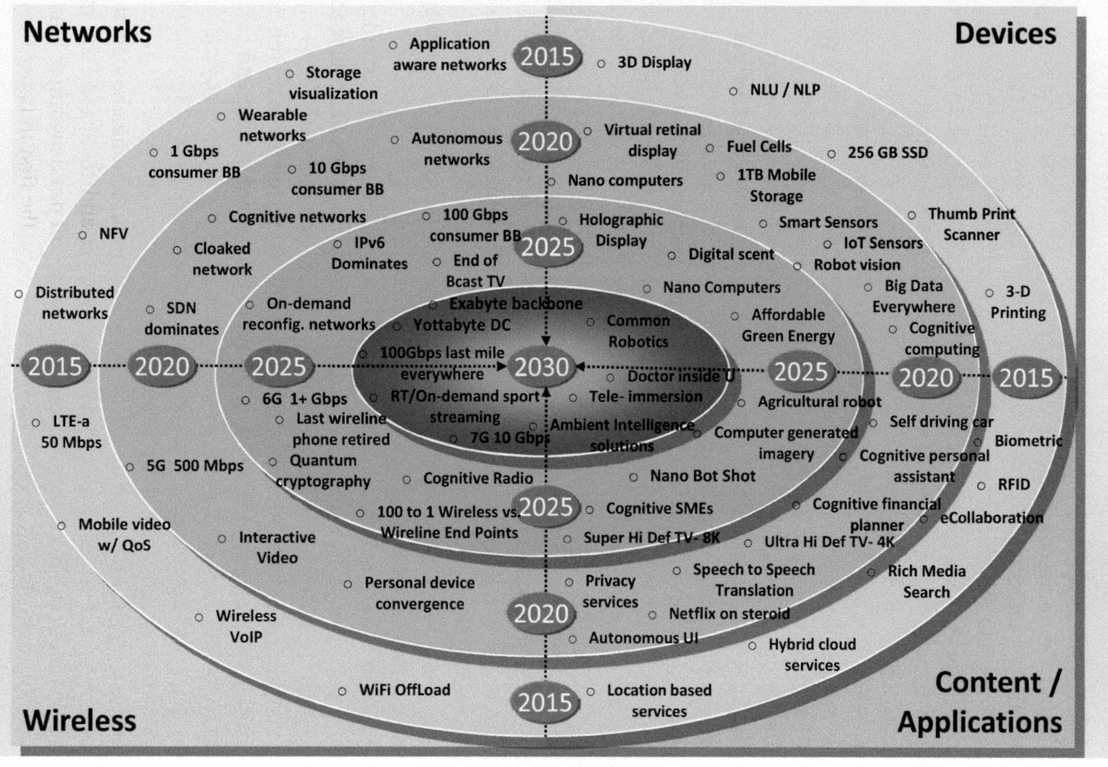

Figure 1.1. What is next?

1.5 The Road to a "Digital Service Provider" Status

Telecommunication companies deliver the digital bits to the customers. There are two kinds of bits: "faster, and faster dumb bits" which is capital intensive with low margins, and/or the "intelligent bits" which has an additional content component, and has a significantly higher margin.

The traditional Communication Service providers (CSPs) have gone through transformation after transformation over the past few decades. All these past transformations have had one thing in common, and that is for the delivery of faster dumb bits, leveraging the technology evolution from analog, to digital, to wireless, to IP. These past transformations have always been about faster connectivity dumb bits, with a higher speed of the last-mile connection.

The next wave of transformations, in an intensely competitive market, will be very different; we call it extreme transformation. In that, the CSPs have to expand their scope of business model and become a Digital Service Provider (DSP) in order to stay relevant. In the DSP world, the customers are not just "human beings", but it also includes billions of sensors and Internet-of-Things (IoT) (Rossman, 2016) devices that will be revolutionizing the digital lifestyle with relevant content enabled by data mining, and leading to decision making, and entertainment. The extreme transformation from a CSP to a DSP status is what we are covering in this book.

The DSP status is a major shift in the business model for CSPs, in that it forces a new focus for the digital lifestyle services and a customer-centric paradigm, to enable the flow of high-margin bits/contents ("smart pipes") with an obsession for QoE (Dai, 2011) in connectivity services, content services, and customer and vendor/partner touchpoints. The high margin-bits must be built upon the high-speed connectivity services, and at a minimum must include video subscription, interactive video such as gaming subscription, and news-and-magazine subscription services, advertising,and continuous introduction of original content, etc.

The DSP status must be enabled by regulatory arrangement for Content purification in order to liberalize CSPs to become responsible DSPs. The absence of such regulation has disastrous consequences as the unregulated DSPs are programmed and prioritized for monetization of end-user engagement, driven by advertising revenue from all possible sources: legal/ethical and illegal/non-ethical.

It was not long ago that we, the authors, were sat around a table enjoying the discussion about our past experiences in transforming massive global tier 1 networks with global footprints which were often headed for a disaster otherwise. In all, we moved the needle to a much higher level for the customers, the shareholder, the employees, and energized all stakeholders with the focus on QoE and the Best-In-the-Industry (BII) practices. We remember our mantra of top three priorities being reliability, reliability and reliability—pick anyone and execute on that flawlessly.

In all, we brought to bear the first network and network services with the BII reliability, the first high-speed IP network, the first optical network with DWDM, the first SDN network with the ability to configure the network in almost real-time to prepare for natural disasters, the first point-and-click provisioning network to enable high-speed connectivity for the business customers in real-time, the first network to support iPhone, then again transformed for number one in IP and integrated services, then again for consolidating networks and OSSs for the number one position for ease-of-use and efficiency in the digitization of the touchpoints and the competitive cost structure for the customers, the first in having implemented flawless operations execution across our global footprints, and on and on.

All CSPs will be going through transformation, time and time again. We asked each other one key question: If we were going to reflect on our experience with transformation, what should be included in a "CSP Transformation" playbook that is timeless and independent of the technology of the day?

We all agreed that the traditional CSPs should do much better than enabling their customers to use their smartphones, laptops and or desktop to get access to the world around us through 100's of disjointed applications. In the new world, we must transform to a DSP, to enable us to go to one portal (voice and gesture-enabled) to get immersed into digital solutions for all our needs. A portal that is built for a passion for QoE, and the integrated value chain of the DSPs from connectivity to digital services. In that, the CSPs must compete with the new and emerging companies with the content and OTT data services (such as Apple, Amazon, Netflix, Google, FB). These recent entrants(compared with 100-year old telephony providers) are promising and delivering differentiated digital services with much higher margins, often delivered over the low margin pipes of the cooperating CSPs. That is a doomsday scenario for the CSPs.

We concluded that, for CSPs to stay relevant,they must compete, not only with differentiation in network reliability for quad-play connectivity services, but also must move up the value chain, and to transform to become a Customer-Centric DSP for digital services. The CSPs, if unwilling to go through this transformation, will face the DSPs who are expanding to include the CSP status and eat the CSP's lunch. In this extreme transformation, the CSPs must:

(1) Redefine the offerings of "connectivity services" to "digital services" and the "touchpoint services"—with end customers and suppliers.

(2) Transform the legacy redundant networks into a unified one.

(3) Redefine the measurements to customer-centric QoE for all digital and connectivity service.

(4) Leverage the timeless Best-In-Industry (BII) processes and practices to ensure an ETE sustainable network performance at a competitively operational efficiency.

(5) Leverage a Service-Over-IP (SoIP) platform to enable the introduction of unified new services with a time-to-market urgency.

(6) Enabled by regulatory arrangement for content purification, which is critical for liberalization of CSPs to become DSPs. The absence of such regulation has disastrous consequences as the unregulated DSPs are programmed and prioritized for monetization of end-user engagement, driven by advertising revenue from all possible sources: legal/ethical and illegal/non-ethical.

(7) Leverage data mining and analytics to know all about the users/end-customers, to anticipate the interests of the individual users, to provide individualized and relevant content, and to monetize the user engagement, and

(8) Provide a unique stream of original content.

This transformation demands an innovative networking platform to make it happen. We know that the technology by itself is never the critical factor for success and has not been in the past, and would not be in the future. After all, most CSPs are purchasing the same types of advanced hardware/software from the same Network Element Providers (NEPs) to build their network with. However, the big league CSP players knowhow to redefine the scope of CSP to become a DSP who knows how to respond to the needs of the customers by leveraging a sound set of processes and BII practices that are timeless and effective. The leaders know how to energize their employees with transformational culture, how to leverage the proven BII processes and practices to piece together the off-the-shelf technology components to build a competitive global network, and how to deploy the proven practices to operate and maintain the network performance, flawlessly.

We have built the differentiated DSP platform, time and time again, and in the process have developed BII practices to keep the transformed network sustainable. This discussion ignited the idea of sharing our overall experience, and specific to the processes and practices, sharing it with the hope that it is a step above and beyond anything anyone ever attempted to write on the subject of CSP transformation.

This book is about what it takes to build,and more importantly, operate a service & networking platform, driven by the needs of its customers. In this book, we explore what the success factors for a national and/or global CSPs are, the critical service flows that the DSP network must be built to support, we explain what the QoE must beset at for the service flows, and we describe what the BII practices are, to build, optimize, and operationalize such a global network to ensure BII and cost-effective QoE for the customers. Also, as critical it is, we explain what the operational disciplines must be to maintain the network at its' top shape at all times.

We did lead successful transformation many times, however, we did make mistakes and learn a lot from our mistakes. The scope of our mistakes and the associated learning is broad and covers not only the CSPs, but also the equipment manufacturers. Some of the key learnings from our transformation programs are as follows:

(1) Must leverage the existing BII practices and processes for the transformation (the subject of this book) rather than building it from scratch.

(2) If your legacy networks and services are NOT performing at the BII QoE performance level, what you need first, is the "optimization" of the legacy CSP networks to achieve BII QoE and SLA (Wieder et al., 2011), before embarking on a transformational program.

(3) Never attempt to build custom solutions for the transformation to a DSP status, but use off-the-shelf technology solutions, built upon a propriety solid and differentiated architectural design of your own.

Based on our collective wisdom with the transformation projects, once your transformation is complete, your end-customers must NOT experience any more improvement in the network and service performance, but must experience competitively priced new services, significant scalability, a much broader base for footprint coverage, and much faster time-to-market for new high margin services.

CHAPTER 2
Preparation for Transformation

2.1 Introduction

In this chapter, and before we dive into the details of transformation, we are highlighting the proven and strategic processes and practices that we have built, over many years, to aid with successful transformations. This chapter is intended to summarize the big picture of what the transformation undertaking must be preparing for. In this section we explore the top-of-the-mind for the transformational changes to come. Often, we use these BII practices as a reference, for the assessment of the CSPs. We often use these practices as a yardstick to gauge the magnitude of the transformation to be undertaken.

During a typical CSP assessment, and leveraging these concepts as a reference point, first and foremost, we examine some of the CSPs' current practices against this reference list. We have seen CSPs all over the map for the use of these BII practices. We have identified low performing CSPs who did not adhere to any of these BII practices, and some who adhered to a few. The more in compliance, the easier and faster to complete a successful transformation time after time.

The transformation is not just a technological change, not a pet-project, and it is not an overnight fix. It is a journey to the higher land. This journey is full of new learnings, setbacks, and re-dos, it needs strong leadership to make it through.

Let us start by sharing our collective experience in the form of a 10-commandment for the preparation of, and a successful transformation:

1. Keep Present Mode of Operation (PMO) (Harmon, 2002) and (Institute Project Management, 2018) legacy networks performing at near ~ 6 Sigma (Tennant, 2001) throughout the journey to the Future Mode of Operation (FMO).

   You must not tolerate the poor service performance of the legacy networks at anytime. For example, we came across CSPs with poor QoE, whose executives were bragging about their upcoming technological renewal projects, aimed at fixing the poor QoE (Laghari and Connelly, 2012) and (Wieder et al., 2011), or their cost structure. The only problem was that they were losing customers to the competition while waiting for these multi-year programs to deliver on the promises.

   In other cases, there were major business customers from the financial institution who were suffering from a loss of revenue due to poor latency in the network, who responded to the CSP's planned technology renewal communication, that "we are going to your competitor, for now, let us know when you are ready for a big-league customer".

As a first rule for transformation, we enforce network optimization, on the legacy networks and services, to ensure BII QoE and adherence to SLA before any attempt for the transformation.

2. Benchmark competition to set the high bar for the transformation.

   Based on our observations, many CSPs did indeed benchmark their competitors, however, in a few cases, we could not see any evidence that their annual goal-setting process was influenced by it. Lack of funding and insufficient resources might have influenced decisions and thought processes.

3. Energize the transformation with a set of bold business goals—becoming the industry leader for customer experience, increasing productivity many folds, slashing the CAPEX unit cost by an order of magnitude, improving the revenue growth above the industry level, etc.

   Based on our experience, there is no place for the Business-As-Usual (BAU) concept. Competition is always moving forward at a rapid pace. This requires decisive action with a compelling set of business goals.

4. Follow the Golden Thread for transformation starting with the PMO (Harmon, 2002), the Master Plan for the FMO, and the Transitional Mode of Operations (TMOs) to the FMO, and benefit realization.

   We did come across a few cases for failed transformation. The effort had created new networks, but had failed to migrate and integrate the legacy networks, and had failed to deliver on the promises for performance and the financial benefits. Migrating a PMO to FMO is a complicated task. It requires a full understanding of the PMO processes, it requires the development of steps (TMOs) and the associated BII processes to move the CSP from the PMO, and through well-planned TMOs to the FMO.

5. Shape the FMO with BII (Best-In-Industry) Practices, e.g., DBOR (database of record), Concept Of One/Zero/None, ASK yourself, Restore First, The3 CP's, etc.

   There is no need to reinvent every component of a transformation puzzle. It takes time to develop new practices and processes. You are doomed to repeat the same shortcomings if you choose to adapt the legacy procedures and not search for, and employ, BII practices. For a successful transformation, it pays to be shameless thieves to bring the best learning and practices from the industry, and to invent a few when needed.

6. Keep the transformation alive and on track with the delivery of In-year ROI (Costantini, 2011).

   Change is tough, it always raises the questions among process associates why to change the PMO. In many cases where transformation was failing, the root cause was the lost energy by the process associates as they were unable to prove the benefits associated with the transformation anytime soon. As a result, they would move back to operate in the well-known and practiced legacy PMO.

As we have said, transformation is a multi-year program. The changes are to be orchestrated and delivered with quantifiable in-year ROI for it to be energizing and rewarding. Keep in mind that the intrinsic value of disruption is failure. We need to accept failures and build a 'failure culture' and learn from it.

7. Engage the Employees and the customers, and communicate with a well-executed communication plan (internal and external) to ensure a sustainable path to FMO.

   We have seen transformation projects being carried in silos. In one silo, a small group of people working on the transformation with little to no communication as to the status and progress, and in a second set of silos, a massive number of process associates, not knowing what was going on or if anything would come out of the first silo. When adding the same disconnect from the customers, and the external partners such as the suppliers, the transformation effort was being treated as a pet-project, and not taken seriously.

8. Keep the progress with transformation visible by a set of metrics.

   The importance of the visibility of the transformation project must not be underestimated. It takes a lot of effort by the process associates to switch over to the TMOs/FMO. It is the fear of the unknown, the fear of disruption, the fear of new learning, etc.

   With a set of metrics to demonstrate the power of transformation, CSPs can energize the entire company to want to, and to compete to switch over to the TMOs and the FMO.

9. Execute the transformation plan Flawlessly, supported by benefit Realization metrics (Institute Project Management, 2018).

   A major reason that many transformation projects failed in their progression, was the inability to prove the benefits associated with the business cases. When the Business Case (BC) (Sheen and Gall, 2015) is being built, many assumptions and promises may present themselves as impossible to keep, over time. However, with a powerful leadership from the CEO down to the Transformation Program Management Office, it is very much possible to deliver on 100% of the projected benefits, on time, at high quality and within budget. One of the key factors for the success of the transformation project is to establish a Program Management Office, with sound leadership abilities, knowledge of the business, and the ability to question and get answers to the "Why Not"questions.

10. Transform the culture for a successful transformation.

   Culture is fundamental to how employees behave. Strong cultures have two common elements: (1) Cultural values: there must be a high level of agreement about what is valued, and (2) Cultural Priorities: a high level of intensity concerning those values.

A successful transformation must endorse the following Cultural Values: Risk-taking, e.g., to project the depth of the iceberg from the top 10% visible above water level, and use that to decide daily; ETE accountability and ETE Process ownership is a must; Golden thread from the business goals, to the ETE process goals, to the organization goals, and the individual goals; Restore First/Repair Later; etc.

The key elements of the Transformational Cultural Priority must be focused on any of the top three below: (1) Results, (2) Results, (3) Results, and the Power of a Vision: A key element of a successful transformation is to inspire the team with the Vision of WHY the transformation is necessary to stay relevant. The Power of a Vision is mostly underestimated.

2.2 Transforming CSP—Why to Transform

Why transform a CSP? Transformation is the most complicated and difficult job. It is a war on the status quo. It impacts the entire company, it upsets the norm, it makes the employees uncomfortable and question why to do it, it makes the customer worried about the impact on the performance of the legacy services, and it makes the CFO question how to pay for it.

You must have a profoundly customer-driven, and financially supported reason for the transformation, and the most essential question is "do we have to transform"?

The first thing we always do is ask ourselves why we need to call for a company-wide transformation. How do we explain it to our employees, our customers, our stakeholders and, most importantly, our shareholders?

The key driver for a transformation must be to save the CSP from a trajectory to a disaster. Failure is not having a competitive advantage in new services, failure is a non-competitive performance of the services and the network, failure is experiencing flat to negative revenue growth, failure is operating a highly non-competitive OPEX and CAPEX platform unit cost, failure is experiencing a much higher churn rate as compared to the competitors, and failure is losing shareholder value. CSPs have to understand the danger of an Innovator's Dilemma, missing the change from sustaining technologies to disruptive technologies.

CSPs are at the forefront of the transformation of the digital lifestyle. In that, if the OTT providers have control of the connections to their portals and their websites to track the customers and to monetize the associated intelligence, the CSPs have control over all access connections to all digital and social media content. This ownership of the access connection must be leveraged as a strategic advantage for monetization of the connections through a variety of business models, including advertising and subscription services.

Based on our experience in transformation, the following shortcomings, one or more, became the driving force for the transformation:

(A) Service portfolio with Low and shrinking margin
(B) Lacking competitive QoE Performance
(C) High CAPEX unit cost to support the traffic growth
(D) Not easy-to-do-business-with (long cycle time for customer-facing transactions, and insufficient touchpoint digitization)
(E) Inefficient OPEX cost structure (long cycle time, high process defects, and lack of automation)

(F) The high cost of "reserve" (driven by defects in the ETE contracting-to-billing processes)

(G) High cost of SLA compliance (Wieder et al., 2011)

Also, it is important to point out that the above shortcomings were the uniting elements for the customers, shareholders, and employees to come together to articulate, with a passionately built financial & QoE logic, the need for the transformation.

The CSPs' business is competitive when it is capable of: (a) BII services and performance,with a broad-based footprint/coverage at the competitive price point; (b) protected network and defensive from cyber-attacks, (c) a network that is scalable and unit cost-effective; and (d) a network that is consistent with the governance obligations in every region of the world in which it is operating.

The transformational triggers for CSPs include *Dated Network & Services vital signs, Poor Business Vital Signs,* and *Non-competitive network/service Vital Signs*.

- **Transformational triggers: Dated Network& Services Vital Sign:**

User Devices, these devices have become more mobile and have become more and more intelligent, multi-functional, capable of supporting all communication and entertainment needs of the users, such as all-in-one TV/Music/Phone/personal assistant/etc., as well as billions of IoT devices are being introduced, such as IoT for smart cities, smart home, smart energy and grid, smart health, smart transportation and mobility, smart factory, and manufacturing, etc.

In the new world, CSPs have no choice, but to provide a network platform for a secure and always-on (with 100-to-1 wireless vs. wired connectivity), with a transparent wireless coverage,throughout the network. The volume and the locality of the content generated/consumed by these devices is unpredictable, but is manageable with a well-thought-out network architecture.

Also, the IoT devices will force the content-processing function to move to the edge of the network, as well as demand high-speed connectivity and symmetric access bandwidth, e.g., to handle video surveillance applications. For example, Given the current CSPs' offering for asymmetric services (DL/UL of 100+ Gbps/10 Mbps), when you activate your cell phone for WiFi calls, the normal fluctuation in the uplink bandwidth forces dropped calls and/or pixelated images, to the extent that a phone/video conversation becomes useless.

Inefficient service-specific and low-speed access, today, the access terminating equipment is custom made for each type of customer-facing access, often offering asymmetric bandwidth, and being housed with their dedicated terminating equipment in a limited space in a Remote Terminal location, with each box having dedicated, inefficient, uplink to the network.

In the new world, CSPs will have no choice but to transform their access infrastructure for significantly higher symmetric bandwidth, while preserving the legacy access protocols. This presents a significant change in the access network architecture, for Multi-Service Access (MSA) (Google, 2005) that must preserve the legacy access as well as the new emerging access technologies, where one unified access terminating equipment is capable of handling all customer-facing access protocols, with a non-blocking architecture, connecting through a common backplane, through an ethernet-based uplink to the network at High-speed of Gbps+.

This architecture must be capable of dynamic allocation of capacity to where the traffic is showing up in the global wireless footprint, as well as the service complexes for transcoding and content distribution.

Inefficient service specific, and many customized aggregation networks, the legacy infrastructure is aggregating low bit rates from the access nodes and the customer sites to the high bit rates for hand-off to the core network for content creation and delivery. This function is performed through a series of inefficient and redundant aggregation network elements for video services, separately for data services, voice services, wireless services, etc. The traffic is often bursty and the peak traffic load is an order-of-magnitude higher than average bandwidth, such as wireless backhaul where installed in the business area, requires less bandwidth at night but the residential areas requiring more of the bandwidth, and the same for enterprise aggregation.

In the new world, the most economical architecture must be based on a unified infrastructure, enabled by SDN (Goransson et al., 2016) dynamic service for all customer-facing services including, wireless/wireline/video, and by enabling the lower bitrate connections to get aggregated in a couple of levels, on ONE unified platform, before delivery to the higher unit cost ports for national/global transport to the content sources and delivery platforms. Unlike the traditional router-centric WideAreaNetwork (WAN) topology, a Software Defined-WideAreaNetwork (SD-WAN) (Butler, 2016) architecture is a new approach to network connectivity designed to fully support applications hosted in on-premise data centers, public or private clouds and Software As A Service (SaaS) (Patterson and Fox, 2014) while delivering the highest levels of application performance. SD-WANs let enterprises and service providers deploy new networking services, applications and features faster, with more flexibility and at a lower cost than ever before. SD-WAN helps network administrators use available bandwidth more efficiently and ensures the highest possible level of performance for critical applications without sacrificing security.

Many redundant core networks, today's CSPs got through aquisitions several core networks, each created to serve a different customer base, some justified in case of a disaster, and often many more are being developed as new networking technologies. There are many examples of core networks delivering IP, TDM, FR/ATM core networks, each presented through a different evolutionary technology from the same vendors. As a result, orchestration and synchronization of service introduction across these networks has become costly, unmanageable, with long cycle time.

In the new world, the core networks have to be transformed to become one unified network, enabled by SDN to handle macro service flows efficiently and reliably, with seamless switch over during a live session, between the access node and/or devices. Also, the unified core must support transit function where wholesale low-featured traffic is being carried. Also, the network support for multiple user personalities has become a must-have and it is paramount that it is most efficiently delivered through one, unified core network.

Congested content management and distribution, today's CSPs often leverage their limited footprint of global Data Centers to collect and deliver content to the users. These data centers are geographically and remotely located and are not QoE effective. Also, there is often a significant traffic congestion on some of the data centers, which renders the QoE useless. Also, the explosion of peer-to-peer multimedia interactions, and the need to transport the same content ubiquitously across all devices, is the dominating theme in content distribution. The idea of a global data center, even if it is relevant, is not sufficient to ensure QoE performance.

In the new world, intelligent content distribution network (Zakas, 2011) and intelligent placement of the content is much needed to transform how content is to be distributed across the global network to ensure QoE performance (CISA, 2019), by enabling an intelligent connection between a user and the source of the content in the global content delivery network.

Multi-Access Edge Computing (MEC) (Sabella et al., 2019) is one such transformational concept, providing distribution of datacenter resources into the mobile network, expected to be a cornerstone of 5G networks. MEC is part of the broad movement of distributed edge computing. The benefits are low latency, high speed, location awareness, improved utilization of RAN, mobile core, and network assets. It enables mobile network operators to gain new data-focused revenue streams and reduce network CAPEX and OPEX costs and it enables application owners and users to improve performance and support new use cases. MEC is driven by time-sensitive and high bandwidth services, including Video, IoT (Serpanos and Wolf, 2017), automation, augmented reality and smart cities.

Long cycle service/applications development, today, CSPs' service introduction is based on a closed-architecture which enables home-grown service development. Often a CSP has a different, non-standard, platform for each service, customized for each access technology, and each new service that requires a custom HW/SW platform to house the services.

In the new world, we need to have an open Service Creation & Delivery framework, supported by a Service Execution Run-Time platform, to enable internet-speed delivery of advanced services utilizing the network for location, preferences, persona, presence, and user info creation.

User/application-unaware Network, the CSPs' old paradigm is based on the fact that the network and the applications/users function as independent entities, e.g., unaware of the location from which the end-user is accessing the network, unaware of the devices being used to access the network, unaware of the type of interaction being requested (e.g., web-based email, multimedia), unaware of what the user's intention is for accessing the network, etc. This thought process is gating the ability of the CSPs to monetize the flow of the bits through their dumb pipes.

In the new world, it is much needed to collect, mine, and monetize user information and intention. The information about the user geographical presence, habits, searches, time and location of the calls and called parties, frequency of calls, etc., this is critical information to use for direct advertising and security. CSPs must deploy infrastructure at the Edge of the network, as well as the sensory infrastructure starting at the end-user device and proximity to the users, throughout the network to enable data collection, mining, and monetization.

- **Transformational Triggers: Poor Business Vital Signs**

We have built the case for transformation, time after time, with the focus on the vital signs for the CSPs. There are two sets of Vital Signs (VS), the “Business VS” and the “Network/Services VS”. One measures the health of the business, and the other measures the health of the Network and Services.

If the score card for the “business VS” is anything close to Figure 2.1, an overall transformation is much needed. This set of “Business VS” provides an excellent base for building a credible business case for the transformation:

Why to transform-Business Vital Signs:

	CSP Business Vital Signs	Score Card
1	Flat ARPU?	●
2	High churn rate relative to competition	●
3	Lack of New / Competitive Services	●
4	Poor QoE Performance	●
5	Miss guided Capex ratio? < 12% or > 18% ?	●
6	High Opex unit cost (Billing DPM >>Network DPM)	●
7	Long cycle time to scale network & services	●
8	Non-competitive pricing	●
9	Non-competitive margins	●
10	High Cost for SLA compliance	●
11	Non-real time customer facing processes	●
12	Lacking adequate world Class trained personnel	●

Figure 2.1. Business Vital signs.

The key Business VSs are outlined as:

(1) **Flat ARPU** (Choudrie and Middleton, 2013): flat ARPU is indicative of a lack of/slow pace for new service introduction. It is a serious problem noted by Wall Street, and reflected in the CSP’s stock price.

Once your network transformation is complete, you have deployed an infrastructure for an open platform for service development, and a “Runtime Execution Environment”, as well as appropriate “Soft Gateways” to support external application providers to deploy new services on your network.

This infrastructure will enable the DSP to rapidly introduce new services/content built on the business model for subscription services and/or advertising.

(2) **High Churn rate relative to competition**: High churn rate is detrimental to the top-line revenue with the customer flight, as well as the high cost of customer acquisition, well in excess of thousands of US dollars per customer.

Based on our experience, when you cut the churn rate from 1.5% to half of that, it would add > 1% to the earnings. If you compare this number with the expected average increase in the earning (~ 3%) of many tier 1 CSPs over the last 5+ years, you will see the power of reduction of the churn rate on the CSPs' financial.

Once your network transformation is complete, your churn rate will be at the low end of the industry, as the DSP will be delivering BII QoE, Competitive Pricing, competitive geo coverage, and trending new services.

(3) **Lack of New/competitive services:** lack of new services has a significant negative impact on the revenue growth and ARPU. The only way to improve ARPU is by injecting new subscription-based services or a price-hike on existing services. Keep in mind that price-hike may not be an option given the competitive landscape.

Once your network transformation is complete, you are on the road to vertical integration with content providers, and the establishment of an ecosystem to enable rapid introduction of new services, as you'll leverage the DSP platform for "Runtime Execution Environment", as well as appropriate "Soft Gateways" to support external application providers to deploy new services on your network.

(4) **PoorQoE Performance**: Customer retention is heavily impacted by the poor QoE performance, as well as competitive service level performance.

Once your network transformation is complete, you will have a DSP platform with BII QoE. (1) a platform that is well architected, and the converged network (wireless/wireline/video) will be driven by QoE for the Macro Service Flows; (2) a well-operated network, driven by the operations disciplines that will be detailed later in this book; (3) a converged access network (for Biz and Consumer) with symmetric bandwidth, and dynamic allocation of access capacity; and (4) a network-aware Intelligent Content Distribution platform.

(5) **CAPEX Ratio not in range:** Quad-play CSPs CAPEX ratio should be in the range of 12–18%. This represents a healthy ratio of CAPEX to gross revenue to ensure that the network is being upgraded and the customer demands are being met with the introduction of improved CAPEX unit cost investment YOY. A low CAPEX ratio may suggest that the network is not built correctly, it is faced with single-point-of-failures (SPOFs) that cause significant customer dissatisfaction, as well as a drain on the OPEX while trying to react to service failures. A high CAPEX ratio is indicative of overspending in the network capacity/new services without adequate monetization.

If the projected ARPU, and the necessary CAPEX investment, in the traffic growth over the next several years, is not supporting the unit cost structure and the CAPEX ratio of the legacy network, then it is overdue for a transformation. A cap-and-grow approach will be required in order to monetize the legacy networks, while placing all the growth on the transformed network.

Let's take the xDSL or PON customers as an example: while the ARPU is remaining flat for those customers, their utilization on the last-mile connection is growing YOY, subsequently, the aggregation and core network have to be expanded to handle the traffic growth. Facts show that the access utilization is growing at 30% YOY, e.g., for an xDSL customer with the utilization of 15%relative to what was bought at the base year. This growth in traffic will force a significant increase in CAPEX expenditure in the access network, the aggregation, the core, the data centers, and the IXPs, without any incremental increase in the ARPU.

Once your network transformation is complete, you will be providing the traffic growth on the BII CAPEX unit cost infrastructure. This will be enabled by a well modeled and understood ETE service CAPEX unit cost, projected traffic volume YOY, and analysis of the trade-off for CAPEX spend on much lower unit cost platforms to support the traffic growth.

(6) **High OPEX (Billing DPM (Tennant, 2001) >>> Network DPM):** High OPEX unit cost, is primarily driven by significant cost of Contra Revenue. It is caused by defects in the quote-to-cash processes.

Once your transformation is complete, your "Reserve" and the cost of SLA will be improving significantly by a factor of 10+, at the same time, the cycle time for customer-facing processes will be driven to near zero.

(7) **Long cycle time to scale network:** Sales and customer demand are unpredictable. The forecast is always wrong and always underestimates the demand. The growth must not be gated by frequent and long cycle time for insertion of costly new technologies, in every step of the way.

Once your network transformation is complete, you have a network architecture, capable of scalability. The demand-driven growth should be supported in ~ real-time for on-net customers, and within days and weeks, (not in months) for off-net customers.

(8,9) **Pricing not competitive/Margins not competitive:** High, non-competitive pricing is indicative of loss of market share and revenue. Also, a shrinking margin is indicative of high CAPEX unit cost, and/or high OPEX unit cost.

Once your network transformation is complete, you have a higher margin that enables the DSP to keep the pricing competitive. This will be possible by improved EBITDA margins, leveraging automation/defect reduction/cycle time reduction, as well as minimized CAPEX by focusing traffic growth on the most cost-effective CAPEX unit cost infrastructure.

(10) **High cost for SLA compliance:** The expenses associated with the SLA compliance must be treated like Billing DPM. The sum of Billing DPM and SLA compliance cost must be treated as "Process Defects".

Once your network transformation is complete, you have improved the performance of network services to match BII QoE, as well as the customer-facing processes (e.g., contracting, ordering, prov., billing, maintenance) to match the network DPM of < ~50.

(11) **Long cycle time with customer/partner facing processes:** The long cycle time is correlated with the number of hand-offs (person-to-person, person-

to-system) and the defects in the process. The number of hand-offs is also indicative of high cost for the OPEX unit, as well as low customer satisfaction in doing business with the CSP.

There are two sets of processes for the CSPs to manage: (1) Customer facing processes (such as ordering, provisioning, billing, trouble management, service restoration), and (2) Network Facing Processes (such as capacity planning, routine maintenance, etc.).

The customer-facing processes are impacting customer satisfaction. These processes are critical for establishing a strategic advantage for the DSPs by establishing a solid case for "easy to do business with". These processes must be driven to deliver in real-time with zero defects and must also be engineered to be self-servicing.

Once these processes are re-engineered/transformed, you must expect a 40% OPEX unit cost reduction for every case of 50% cycle time reduction, and significant progress in the digitization of the touch points, e.g., customer/partner web-enabled services, delivering self-servicing transactions at zero cycle time.

(12) **Lacking adequate World-class trained personnel**: This is one of the most pressing issues when going through a transformation. How do we get the trained personnel to care for the new platform and business processes?

Once your transformation is started, you are on the path to getting a network of champions (from your own CSP's associates) fully invested in the transformation, so they can coach others on how to use the transformational tools and procedures to create transformational benefit to the bottom line. This group of transformational individuals should replicate the organization and include your star performers. It is key to not just pick the geeks who are interested in the technologies. You want people who can work horizontally across the organizations and who have excellent communication and networking skills. It's most important to not have the early adopters adopt, but that influencers adopt. Getting those folks on board early is critical to train the workforce.

- **Transformational Triggers: non-competitive network/service Vital Signs**

Today, the CSPs are being benchmarked for network performance in the six network performance categories (RootMetrics, 2020): call performance, network speed, text performance, data performance, overall performance, and network reliability.

This approach is dated, not focusing on the critical Service Flows, not measuring the QoE during the peak traffic time,and is underestimating the performance shortcoming by comparing the best and the worst performers, only apart, by a couple of percentage points while the customers of the poor performers are suffering from poor QoE performance, especially during the peak traffic time.

It is our assessment that, over time, the benchmark is rapidly changing to include the QoE performance (Mellouk and Mushtaq, 2017) of the 7 Macro Service flows (MSF) during the peak traffic time: browsing, streaming, Interactive Video (e.g., gaming), VoIP, access bandwidth (consumer & Biz), and digital service availability.

It is important to note that a drill-down of the "QoE performance" is needed to establish the root causes for poor performance. This drill-down capability is critical to ensure that the optimization efforts of the legacy networks are laser-sharp focused to deliver to BII QoE.

If the current customers of the CSPs are experiencing QoE as depicted in Figure 2.2 during the peak traffic hour, while the network is enduring many inherent network failures daily, you need to focus on "network & Services optimization" of your legacy networks as a first step before the transformation.

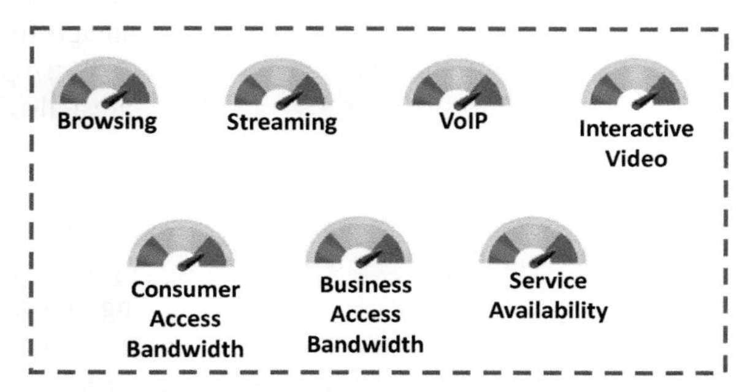

Figure 2.2. Is this the CSPs' peak traffic QoE under failures?

Your legacy networks must always function to deliver BII QoE, at a predictable cost for the SLA compliance. The green/yellow/red indicators are based on the threshold for the expectation for customer experience, as we have covered the details later in the book.

BII QoE performance must be delivered at all times, independent of any planned network improvement and/or the network technology you use in your legacy networks.

Once your network & service optimization on the legacy networks is complete, you should be delivering the QoE performance, for at least 4 of the above 7 MSFs, as outlined in Figure 2.3, during peak traffic time and in a rainy-day (w/failures) scenario in your legacy networks. The key is to ensure BII performance, in every legacy network, including wireless/wirelines/video networks, under well defined failures:

The Network & Services optimization of your legacy networks, once done right, will move the needle to the "Green" status for the following service flows: Browsing, Streaming, VoIP, and Service Availability. This requires two specific optimization activities to get there: (a) Architecture Optimization, to improve the resiliency of the legacy networks, and (b) Operations Optimization, to ensure flawless execution of the Operations Discipline to ensure that the integrity of the Architectural Optimization is maintained, as will be covered later.

To move the needle for the other service flows (scalability, Bandwidth, Interactive Video and latency-sensitive applications), a transformation of the network infrastructure and processes is needed.

In our assessment of many tier 1 networks, we could verify the BII performance in a sunny-day operation, meaning under no network failures (which is seldom the case), but under failures and/or during peak traffic, the QoE was always less than

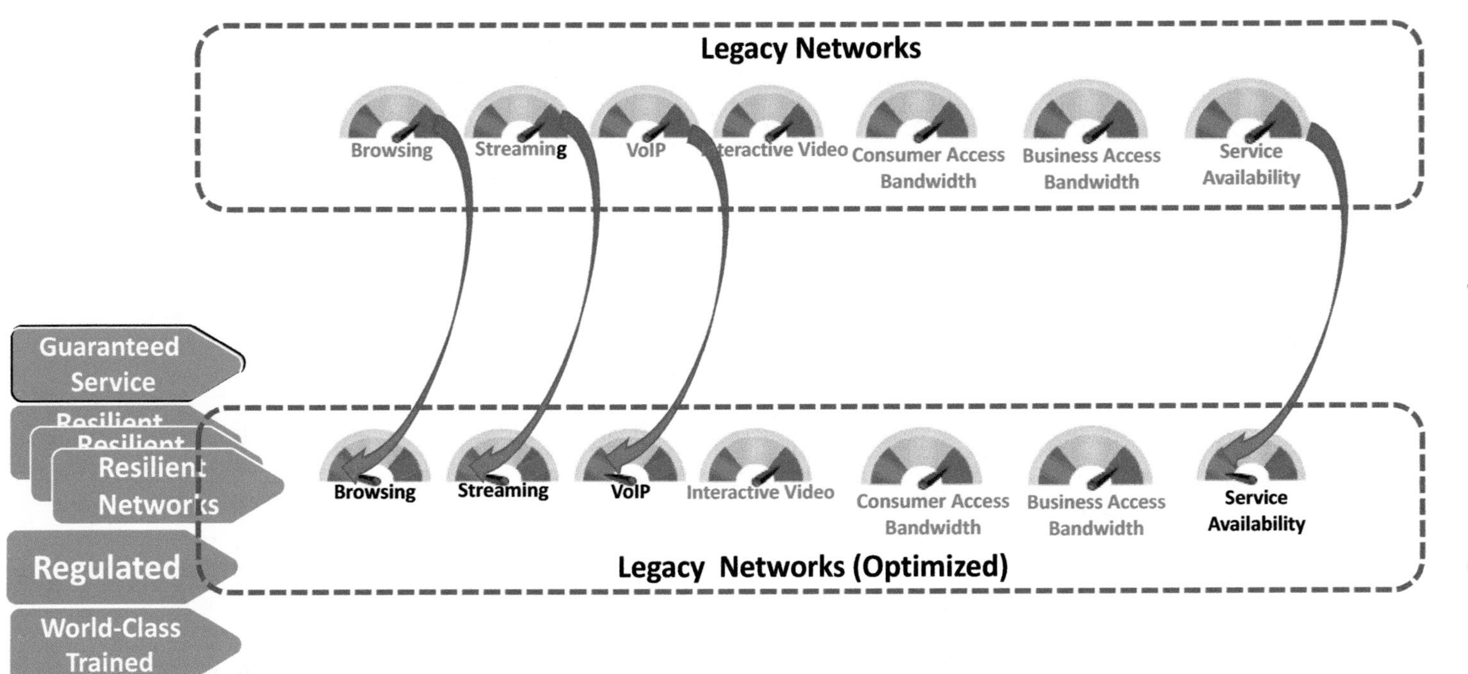

Figure 2.3. Optimized legacy networks-minimum 4 of 7 MSFs turn green.

desired. The global networks are easily experiencing 100's of major failures weekly. What differentiates a resilient network from a "me-to" network, is the resiliency of the network to perform and to deliver to the BII QoE under failures and during peak traffic time.

When we are assessing the legacy networks, we look for variation in the network KPIs under failures (Marr, 2012). Keep in mind that variation must be recorded during peak traffic time. Averaging the KPI will wrongfully lead you to believe that there is less-of or no QoE problem. As a part of probing the network, it is wise to start failing the live network one piece at a time, during peak traffic hours, to test its resiliency architecture, such as: failing core routing complex, failing core transport spans, failing aggregation nodes, failing data centers and service nodes, failing CDN (Zakas, 2011) nodes, access nodes, last-mile connectivity, etc. If you see a hesitation in your network operations to fail the network during the peak hours, it is very telling for the questionable resiliency of your network.

We have assessed many legacy networks, with unbelievable and often disappointing results: we could see, in real-time, massive variation in the QoE under failure. We could see disruption in service across geographical areas, ETE latency shooting up to 400+ ms (CISA, 2019) which destroyed VoIP and made the browsing experience unacceptable with a performance of more than 25 seconds to download a typical web page (~ 9 MB), the gaming experience was also impossible. Furthermore, we could see a major uptick in the on-demand maintenance, which affects the customer and is extremely inefficient to fix in real-time, etc.

To fix these QoE problems, it takes a thoughtful *Architectural Optimization* to fix the legacy network for a rainy-day operation. It also takes a massive *Operational Discipline* to ensure that the network is guarded with flawless execution and performance, and it takes superb *Real-time Dashboards* to proactively and predictably identify the failures and to marshal resources to fix the problems before they impact the customer.

Once you achieve the BII service performance under failure on your legacy networks, you are ready to embark on network transformation to address the rest of the "RedScores" in the table above (e.g., through unification of network infrastructure, scalability, high-speed access, Footprint expansion, and introduction of delay-sensitive IV applications). The key in this step is to transform the network architecture, to introduce a new service platform, and to redefine the last mile access bandwidth.

Once your network transformation is complete, you should be delivering scalability (i.e., unleashing sales) and new services, highlighted by the blue color in Figure 2.4.

A Successful transformation must be aided by key enablers. The *key enablers/ tools* for CSP transformation can be summed up as follows:

(1) *Best-in-Industry Practices*...to introduce and to institutionalize the practices for a Best-In-Industry DSP.

(2) *Operations Discipline*...to ensure flawless execution of daily operational routines.

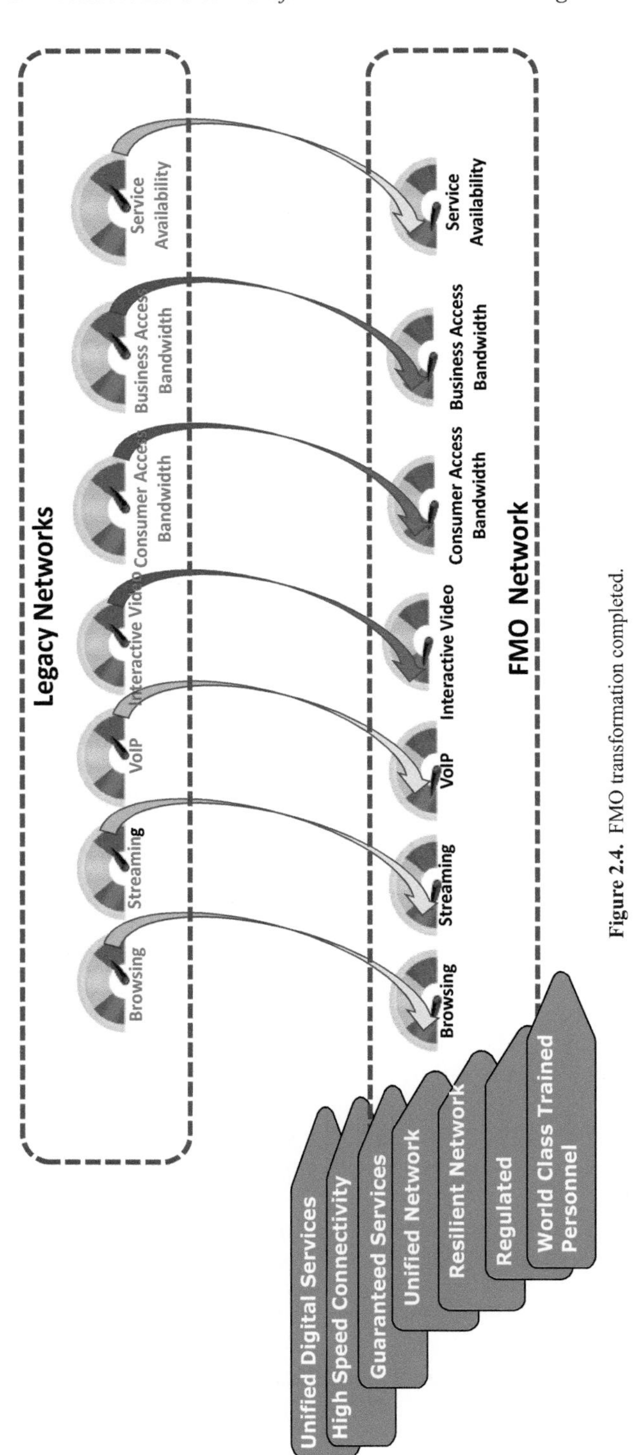

Figure 2.4. FMO transformation completed.

(3) *Disruptive technologies* such as SoIP platform, IoT sensors, IP, SDN, NFV, ICDSs. These are instrumental in introducing new services, improving customer experience, and reducing CAPEX unit cost.

(4) The *technologies that enable "Convergence"* of services/networks (such as IP, a common and open service introduction architecture such as SoIP), as it is essential for convergence of wire line, wireless, video, and to significantly reduce the unit cost through Concept of 1, Concept of 0, Concept of None…. these will be explained in this book.

(5) *Extreme Automation and business process transformation*…automation that is focused on zero "defects" and "cycle time" to be effective in transformation. The focus on "defects" and "cycle time" in every aspect of the CSP processes is the driver for superiority in customer experience and OPEX unit cost reduction.

(6) *New competitive services* often call for transformation, these services could include, Original Content, Intelligent Personal Assistant (IPA), IoT, etc., as they provide the engine for revenue growth, and

(7) *Regulatory relief* (Benjamin and Speta, 2019) and the concept of Fare Market-forces is another key enabler for transformation, as it often drives unit cost reduction for both CAPEX and OPEX.

The key focus of the CSPs over the next 5 years are: (1) slash operational unit costs throughout the scope as outlined above, (2) minimize CAPEX unit cost, (3) cut commercial costs, (4) significantly Enhance ARPU through the introduction of new content-based services, and (5) expand the service footprint to increase market share. Needless to say that Network Quality will continue to be a strategic differentiator going forward.

The network must be transformed, and the vertical integration with content generators will be the wave of the future for CSPs to help them to become DSPs. The convergence of devices, networks, content, and IT has led to the development of new services that is driving the demand for more bandwidth, as well as significant competition from OTT service providers.

2.3 Transforming CSP—Scope of Transformation

Transformation is a massive and difficult undertaking which starts with the understanding and quantification of the performance of the Present Mode of Operations (PMO), customer expectations, and the competitive pressures; then, for a competitive posture, one must formulate a Future Mode of Operation (FMO), driven by the competitive business targets for the future state; and finally, it is expanded by the Transitional Mode of Operations (TMOs) to bridge the gap from PMO to FMO in logical and economical steps to ensure that the transformation is paid for and is CFO proof, mostly through in-year ROIC; the end result is a seamless migration of the customers to the new platform. It is critical to note that the legacy CSP network must be continuously optimized to provide BII QoE at all times, and throughout the transformation and the migration process.

The scope of the transformation must cover, at minimum, the 5 key domains and it's 55 components, as presented in the picture below (Figure 4.5):

i. Network & Services
ii. Network Processes
iii. Network Management
iv. Operations Discipline
v. Program Management (key focus on the realization of the transformation benefits, exit criteria: a program is exited when it is completed or transitioned to operational processes)

(i) Network & Services is comprised of (a) connectivity and digital Services (content, SoIP, news/magazines, HIS, High-speed Access, Wireless/WiFi, Security, CDN, Data, IPTV, Wireline voice, etc.); (b) Network (unified Core, unified Access, unified Aggregation, Regional, CDN, Common Backbone, unified transport, unified IXP, USDC, UIDC, Security), and (c) Management applications (Performance & Quality, Provisioning, Capacity Management, Maintenance, and Care, etc.); (ii) Network processes are comprised of Customer-facing, Network facing, Partner facing; (iii) Network management (Fault management, Performance management, Fault correlation, Asset discovery, Transmission management, and traffic analysis); (IV) Operations Discipline (Customer communication, Ask yourself, Network events, 3CP/MCB/TCB, Root Access, Outage management, M&Ps, Disaster Recovery); and (v) Program management (T-Plan, Q-Plan, Benefit realization, Gate control, e.g., checklist, certification).

Figure 2.5 is a representation of the scope of a typical transformation:

2.4 Transforming CSP: Best-In-Industry Practices

Within the context of this book, the "Best-in-Industry practices" is associated with the proven practices, developed by the practitioners, researchers, and decision-makers with executive accountability from Tier 1 CSPs/DSPs that have established the methods and procedures that we call "Best-in-Industry Practices". These BII practices have contributed to the bottom line of the CSPs in a measurable and differentiated way, and have been instrumental to establish few CSPs as BII CSP and DSPs.

We have made sure to accumulate a set of practices, collectively, that have contributed to the competitiveness and sustainability of many of the global CSPs instrumental to the delivery of the results, as outlined below:

- ❑ Connectivity & digital services:
  - ➢ Revenue:
    - ✓ Up to 20% increase every other year due to new Digital Services
    - ✓ Up to 10% increase every other year due to footprint expansion

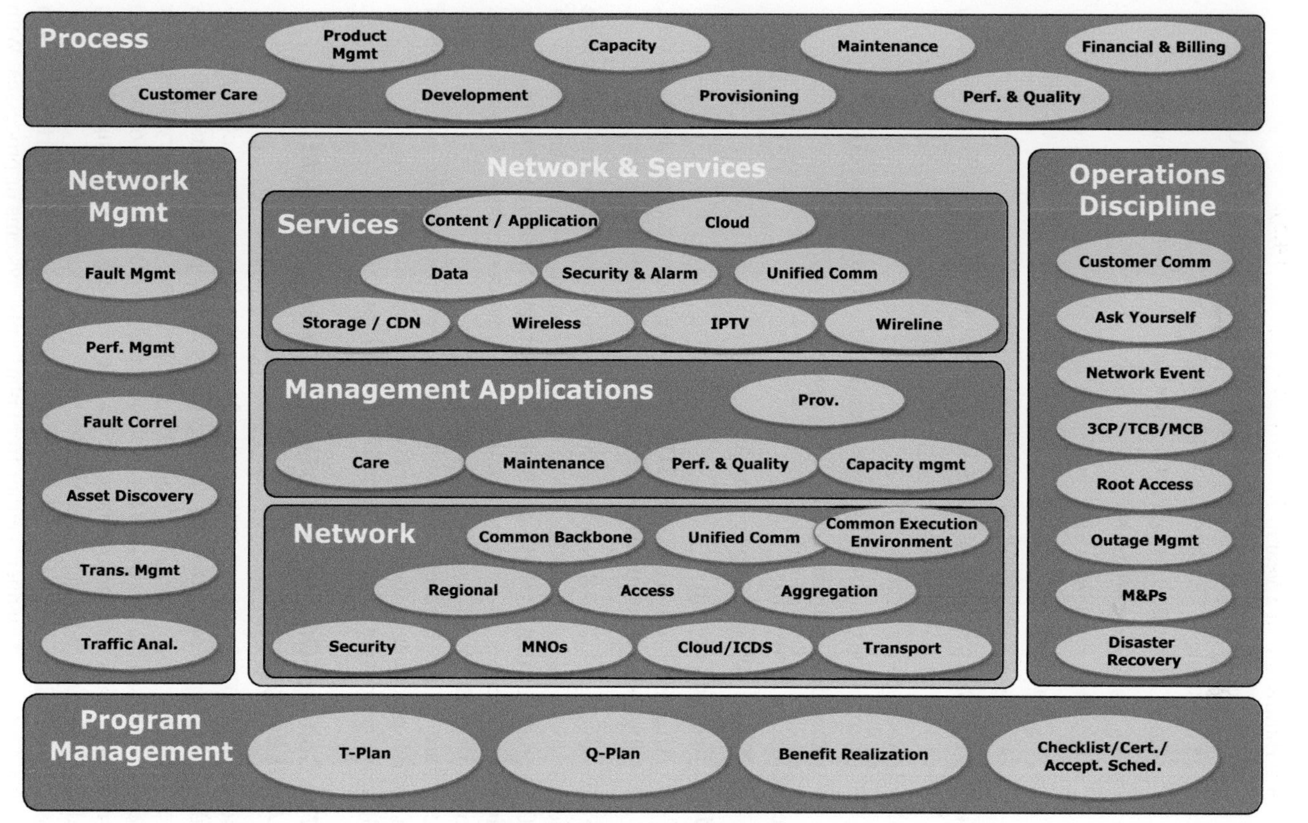

Figure 2.5. Scope of transformation.

- Performance:
  - ✓ Bandwidth in 5 years:
    - Consumer: 100++ Mbps symmetric, with bursting to Gbps
    - Business: On-demand > 40+ Gbps symmetric
  - ✓ QoE Driven Performance Targets (5000 Kilometers network):
    - Latency < 80 ms RTT
    - Jitter < 10 ms
    - Packet Loss < 0.01 %
    - Service Availability (Biz) > 99.999%

- ❑ Capital unit cost reduction (ETE services for wireless/wireline/and video) > 70%
- ❑ OPEX Productivity target > 80%
  - ✓ Zero cycle time for customer-facing self-servicing processes
  - ✓ Zero defects for network-facing processes
- ❑ Scalable ETE Network:
  - ✓ Scale: 100's of Tbps per node
- ❑ Macro Service Catalog:
  - ✓ High Resolution (4K, 8K, 3D) Interactive Video, Streaming, Browsing, High-speed symmetric Access, Service Availability

We have leveraged the BII practices in the construction of the FMO which were instrumental to the delivery of the above business results. Here is the preview of some of the practices that we have applied to many transformations throughout our professional experiences. We will outline these BII practices in this chapter, namely, Co1, Co0, CoN, DBOR, AYS, 3CP, SDR, etc.…

The picture below (Figure 2.6) presents a bird's-eye view of the application of some of the BII practices to the overall network architecture:

Below, we are providing a highlight for these BII practices:

2.4.1 The Concept-Of-One (BII-Co1): Do It Once, Do It Right, and Use It Everywhere

The essence of the Co1 (at&t, 2003) is to reduce costs and to create efficiency by consolidating multiple organizations, networks, systems, platforms and processes into one. With the Co1, you do it once, do it right, and use it everywhere. For example, this Co1 was the key driver for the consolidation of many networks into one unified network, consolidation of many OSSs into one (e.g., one billing platform), consolidation of many redundant organizations into one, etc.

The basic principle of the Co1 is that we will leverage the scale to perform a given function in a centralized manner—either from platforms, systems or processes. With the Co1, we will not have multiple organizations, platforms or systems performing the same function. Instead, there will be clear accountability and responsibility for performing any functions.

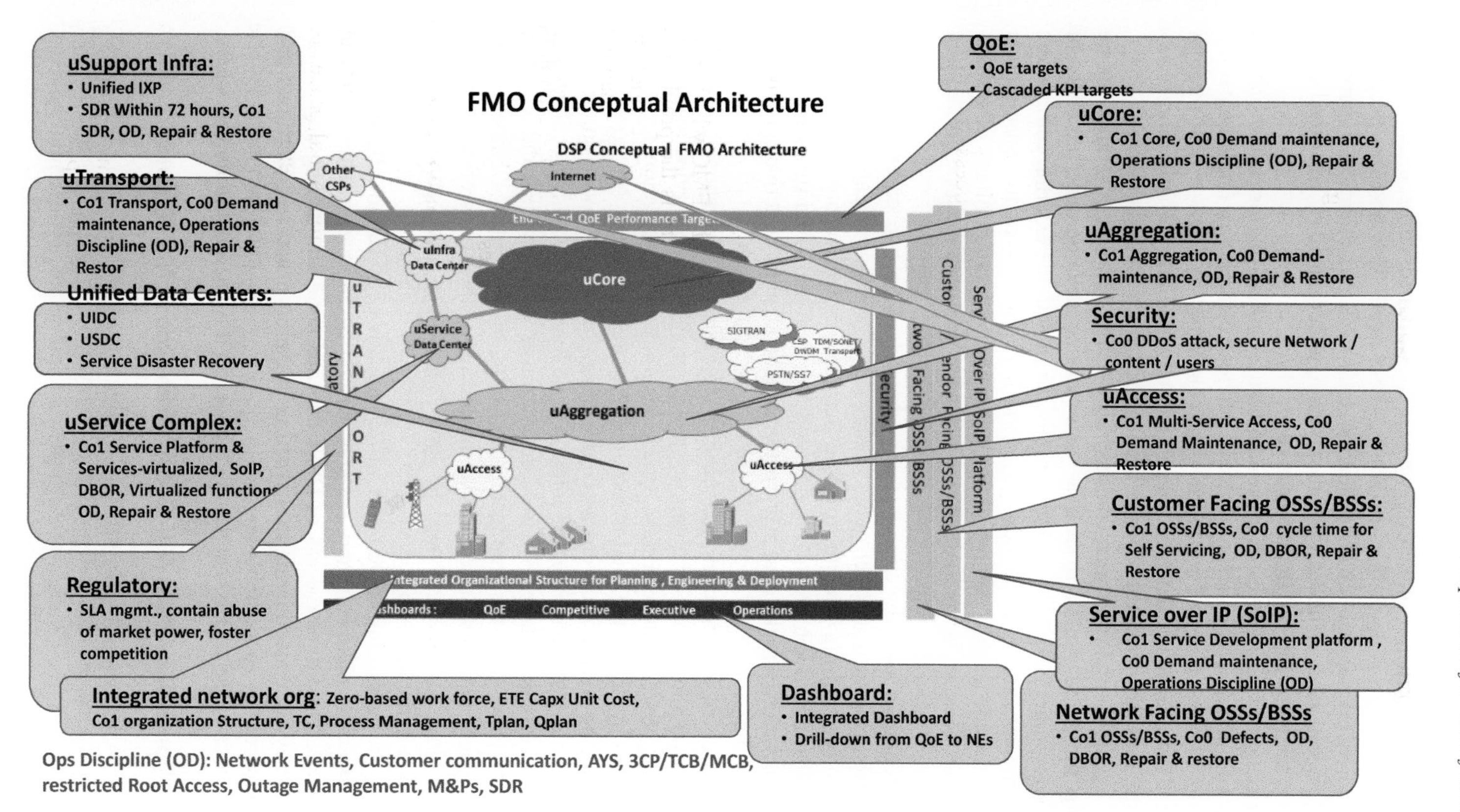

Figure 2.6. A birds-eye view of the application of the BII practices to the overall DSP architecture.

To achieve the principle of the Co1, we need to be aligned with a common vision and plan of record to support that specific business need. This business need can vary from a network aggregation consolidation, to an OSS consolidation, to network consolidation initiatives. Without a common vision, underpinned by an agreed-upon architecture, the team will fail in its objective.

Describing the Co1 in terms of consolidation to common platforms and systems is a fairly well-understood concept—but with obvious difficulty in implementation. The benefits of these types of consolidations include not only lower cost through efficiencies, but also fewer database integrity issues, less fall out, and more flow-through and, thus, achievement of faster time to market for the customers.

For example, multiple organizations, e.g., Labs, IT, Sales, should not be performing the same function, such as new service development, in different organizations—competing with each other rather than with the external competitors to win customers; multiple redundant systems lend themselves to consolidation under Co1; introduction of a common DBOR Bus eliminates multiplicity of system-to-system interfaces; so is the multiplicity of legacy networks must be unified into one DSP network, etc.

The meeting of our objective is rarely dependent on one organization and function for its achievement. Much of the work requires cross-organizational teams—most frequently requiring Architecture, the Service Realization organization, and frequently, Research. Each organization provides critical skills and functions to the support of the initiative or the goal. All are needed, and all must be active participants in the work of the team.

Participation in a team does not mean a "free-for-all" in responsibility. The architecture team should not be doing research; nor should the service realization team be doing systems development. Each function or organization must have clear accountability under the total umbrella of the overall project. Any given work function should be performed in one and only one organization to avoid duplication.

This does not mean that only one solution or architecture should be evaluated or pursued in the formative stages. We need innovation as well as critical and objective differences of views until the architecture and solution is adopted. Competition in the earliest stages is healthy to ensure we have thoroughly and carefully analyzed alternatives. Once selected, we need to unify beyond the single architecture and solution to win in the marketplace.

With the business imperative to be cost-competitive, we cannot afford to have multiple processes, platforms and systems with their unique requirements and resource needs. The future success of DSPs depends upon eliminating duplication and redundancy by consolidating and creating common processes, platforms and systems supported by accountable teams.

The essence of the Co1 is that you do it once, do it right, and use it everywhere.

2.4.2 The Concept-of-Zero (BII-Co0): When One is Not Enough

The essence of the Concept-of-Zero (at&t, 2003) is zero-touch, e.g., self-service processes. This concept is focused on eliminating two evils of cycle time and defects. In its simplest form, the Concept-of-Zero strives to eliminate non-essential systems, functions, or work. However, the real power of the Co0 comes in challenging traditional assumptions and replacing them with a new set of paradigm-breaking or "extreme" assumptions that radically change the problem set.

As we said before, the Concept-of-One is a powerful tool for reducing costs and creating efficiency by consolidating multiple organizations, systems, platforms and processes into one. With the Concept-of-One, you do it once, do it right, and use it everywhere. However, certain circumstances require an even more powerful tool—the Concept-of-Zero. This concept is pushing for extreme automation with the goal of Zero defects and Zero cycle time.

For example, with this Co0, when we applied it to the maintenance process, we eliminated on-demand maintenance activities, rather than attempting to optimize them. This approach required a radical shift in the design paradigms where solutions were engineered with enough excess capacity and redundancy that services could be rapidly restored by automatically shifting to spare equipment and the failed equipment replaced as a routine maintenance event, likely performed by the vendors. As a result, customer satisfaction was improved significantly, and operational expenses were replaced by a small amount of capital expenditure; a transformation to an auto-inventory network is transformative for zero defects in the accuracy of the network assets which is critical for reuse, as well as e-enabled self-servicing for customers and the supply chain; system retirements to zero; flow-through servicing; retirement of the legacy network elements to zero; etc.

Applying the Co0, while on the surface Co0 deals with elimination, its actual power comes from strongly challenging the status quo and examining extreme or paradigm shattering assumptions that would radically change the function or activity. In many cases, it is more productive to construct an extreme view of the future state and work backward to the present situation than it is to try to move forward from the present state and its current complexity, as described above in terms of the maintenance process.

In many ways, the Co0 is like Root Cause Analysis in that it requires a sequence of "why" questions that challenge basic assumptions and go on asking "why" until the most fundamental drivers or causes of the activity are discovered. Radically changing these fundamental drivers can create a paradigm-shattering effect that causes the whole problem to be viewed differently. The capital/operating expense tradeoff in the maintenance process is an excellent example, illustrative of the creation of a resilient network where failures are meditated by switch over to the redundant capacity.

Another example of the paradigm-shattering effect of the Co0 can be found in how it is being used to drive the CAPEX associated with the "Running the Business" (RTB) expense to zero. While zero RTB expense is indeed an extreme assumption, and on the surface appears ludicrous, the exercise pushes the envelope on what is achievable much more than a target of an incremental reduction of RTB by 10 or 20 percent would ever achieve. Moreover, the series of "why" questions is exposing and challenging the fundamental RTB cost drivers. It is important when applying the Co0 to focus on truly extreme or paradigm shattering assumption sets.

As the telecom industry continues to reshape itself in fundamental ways, the advantage will accrue to the CSPs that can get ahead of the financial trajectories of the industry by radically reshaping their cost structure. The Co0 is a powerful tool for achieving that objective.

2.4.3 The Concept of None (BII-CoN)

The essence of the CoN (at&t, 2003) is the "Dark Factory Automation". It is the extreme case of Co0, when applied to all customer-facing processes, such as service provisioning, it transforms those processes into self-servicing processes. This concept is focused on eliminating the two evils of cycle time and defects from the customer-facing processes. It is the most powerful concept to drive customer satisfaction, and to enable the customers to be in control of their needs from the CSPs.

2.4.4 Database of Record (BII-DBOR)

CSPs' systems architecture is comprised of thousands of OSSs/BSSs, most with a dedicated and local view of databases and data dictionaries. This situation is challenged with much more complexity associated with the acquisitions and mergers over time. As a result, data accuracy is significantly challenged from sales to billing, contributing to excessive contra revenue.

We built the concept of DBOR (data warehouse-Database of Record) (at&t, 2003) to fix the data accuracy from sales-to-billing to ~ 100% accuracy level, with a focus on a single customer view of all services in use. With DBOR we interfaced all legacy OSSs/BSSs to present a single customer view through DBOR under strict data dictionary management for sanctioned source of data elements. We developed a data cleansing apparatus to clean the data at the source, and we developed a data migration platform to get the data from the source to the DBOR.

DBOR became a critical customer-centric platform to reduce contra revenue, and to enable flow-through provisioning, becoming an enabler for the CoN for customer-facing processing.

2.4.5 Ask Your Self (BII-ASY)

CSPs' technicians routinely touch the live network for routine, and demand-based maintenance, as well as a variety of capacity insertion, or provisioning activities daily. The volume of these activities is in the range of thousands daily for a tier 1 CSP. A simple mistake in carrying out this work was impacting the live network, and often were customer impacting.

The concept of AYS (at&t, 2000) is aimed at dramatic reduction (> 99% reduction) in Plant Operating Errors (POE) across the global footprint of the CSPs.

AYS principle was first developed to ensure a standard and predictable outcome for touching the live network by hundreds of thousands of technicians daily.

We developed and institutionalized a process to achieve this objective. In that, we developed a questionnaire and the associated processes and systems to ensure the technicians who are touching the live network are doing so under process management and control.

We developed this concept for the technicians in Network Operations, and soon expanded that to other parts of the business, including network planning and engineering.

For example, in network operations, this concept was developed fully, with the associated processes and systems. We mandated the execution with across the board training, and put the appropriate process in place to ensure that all technicians do comply with it. Under AYS, we rewarded the technicians asking these specific questions, and have the answers ready, before touching the live network:

(1) Is the M&P (Methods and Procedures) developed for this maintenance work?

(2) Is the M&P approved by the supervisor?

(3) Is the technician/vendor trained to execute this M&P?

(4) Does the M&P have coverage for the back-out procedures, in case of any unpredictable service interruption?

(5) If the vendor is executing the M&P, is there a responsible technician available to monitor the vendor?

(6) Is this work scheduled to be performed during the authorized time?

(7) Is NOC (Network Operations Center) informed and scheduled to monitor the execution of this work?

(8) Are there any other related maintenance activities scheduled at the same time in the same office that may interfere with this work?

It is key to note that there must be processes established to ensure that each step is carried out in full compliance with the standards for the AYS. With full implementation of the concept of AYS, we delivered a dramatic reduction in POE across the global footprint of many CSPs.

2.4.6 Outage Management (BII-OM)

The concept of Outage Management was developed, to get activated in real-time with any incidents, and operated in parallel to complete the root cause analysis while the fixes were being developed and deployed.

CSP technicians routinely touch the live network for routine and demand-based maintenance, as well as a variety of capacity insertion, or provisioning activities daily. The volume of these activities is in the range of thousands daily. A simple mistake in carrying out this work will impact the live network and often the customer.

The extreme analysis of the root causes indicated that the processes and M&Ps that the technicians were following were not updated with the learnings from the past failures. For example, the correction to a failed work step was not updated in the procedure, as a result, the same failed procedure was repeated again and again by other technicians.

Under Outage Management, we ensured that the fixes were root caused in real-time, and were being institutionalized in the day-to-day processes. This was aimed at ensuring that the same process defect does not happen a second time.

We learned that it was extremely hard to start this Outage Management activity and to staff it properly during the initial deployment as we were experiencing an alarming number of incidents, however, it proved quite effective and manageable over a very short time frame, as the institutionalization of the learning drove the defects to a bare minimum.

2.4.7 Communication, Command, and Control Process (BII-3CP)

3CP (at&t, 1999) was developed to manage network anomalies. Network anomalies are defined as any impacts or potential impacts on business customers' and consumers' outages on a 24/7 basis.

The principal mission of 3CP is to eliminate or reduce customer impact, protect the CSP's brand name, and reduce or eliminate regulatory reportable network incidents. 3CP is designed to:

1. Effectively manage the communication of information and data within CSP when there are incidents, or disasters that affect or have the potential (such as an emerging hurricane) to affect the CSP services.
2. Respond in real-time to incidents that affect CSP services and network, equipment and/or people.
3. Manage service restoration efforts as efficiently and expeditiously as possible until there is no longer a threat to the network.
4. Establish treatment of redundant equipment failures with the same priority as service impacting outage.

3CP starts when the service impact to the customers reaches a specific Network Performance Level (NPL) threshold and/or failure of a redundant equipment. Different levels of NPL were introduced with a specific threshold to trigger the 3CP activities.

During a network incident, 3CP provides for:

1. The timely assessment of potential or developing crises or disasters that could impact service to the CSP customers or network equipment.
2. An emergency communication process to notify the appropriate organizations that could be involved in the incident.
3. The establishment of Technical Control and Working Telecom Bridges (TCB) so subject matter experts and field personnel can efficiently and appropriately share pertinent technical information concerning the restoration options in real-time.
4. The establishment of the Management Control Bridge (MCB) comprised of a leadership team with the responsibility to determine restoration choices.

Three tiers allow for the flow of communication throughout the CSP. Three-Tiered Communication Structure:

a. *Tier I.* Responsibilities include analyzing technical problems, implementing tactical plans, and *activating working bridges* as appropriate. Tier I activation criteria is NPL-1.

b. *Tier II*. The CSP NOC *Technical Control Bridge* (TCB) must be activated to address network incidents and/or hazardous conditions before they reach MCB activation criteria. TCB activation criteria are NPL-2.

c. *Tier III* is the MCB itself. The *MCB is activated* when specific criteria are met. The primary contacts are referred to as the Duty Officers (DO). These are Tier 3 Engineering/Network Operations Directors. MCB activation criteria are NPL-3.

The Core Team is accountable for the overall network recovery effort. Depending on the nature of the incident, the NOC Duty Officer will request the designated representatives from other organizations to join the MCB as needed in real-time.

2.4.8 Closed-Loop Performance Management System (BII-Integrated Dashboard)

We created this concept to ensure that the interests of 3 groups of people are being integrated into a culture for winning in the market place, where there are millions of customers, hundreds of thousands of employees, and millions of shareholders.

This culture for winning had to bring laser-sharp visibility into the performance of the use of the CSP's services, had to enable and decentralize decision making by the employees to care for the customers in real-time, and had to reassure the shareholders that their investment is secure and rewarding.

As a result, we developed an integrated closed-loop set of dashboards comprised of:

- QoE (enables customer-centric and fact-based view of customer experience)
- Competitive (ensures competitive posture)
- Executive (enables executive oversight)
- Operations management (provides real-time management engagement when needed)
- Working-level (enables real-time actions, affecting customer service and CAPEX)

The BII practice stems from the ability to make these dashboards operate in real-time, and integrate these dashboards with drill-down capability for fast and fact-based decision making.

2.4.9 War Room-Cross-Organizational Issue Resolution (BII-WR)

We developed the concept of War Room to attack the critical customer-facing issues that were caused by the presence of the siloed organizations and systems which were operating independently on an ETE process with many hand-offs, long cycle time, and customer-facing defects. The objective was to quickly identify the root causes, establish temporary/quick process fixes to stop the bleeding, and over a short period, to reengineer the ETE process and systems as a part of the on-going improvement. Here is the organisation of a typical War Room:

A typical daily Executive-led War Room:

1. The timely assessment of potential or a developing crises or disasters that could impact customer-facing services, e.g., Customer Billing Error as defined by Billing DPM and Contra Revenue.
2. Establishment of a designated executive lead for the War Room.
3. An emergency communication process, led by the War Room lead, to notify the appropriate organizations' leadership to be involved in the War Room.
4. Establishment of empowered SMEs from respective organizations to attend the War Room on a daily and as-needed basis.
5. War Room exit criteria: stop the war room based on 80/20 rule, i.e., when 80% of the root causes are identified and quick process fixes are in place. For example, for a major business customer with a significant billing error, through the War Room, must identify the billing defects and the root causes: 40% in the contracting processes, 30% in the Ordering and provisioning processes, 20% in the late disconnect process, and 10% others, and must develop and deploy quick process fixes.
6. Frequent read-out on the identification of the root causes, quick fixes, deployment status of quick fixes, and reengineering of the ETE process, and the progress on the KPIs, e.g., Billing DPM.

2.4.10 Restore First, and Repair Next (BII-Restore & Repair)

3CP is aimed, first and foremost, at restoring the services, being impacted by network anomalies, and then repairing the root cause of the service outage. As explained before, the BII-Restore & Repair practice is executed when the service impact to the customers reaches a specific Network Performance Level (NPL) threshold under the BII-3CP.

However, we wanted the principle of "restore first, and repair next" to be the standard mode of operations for all service impacting incidents, below the NPL threshold.

With the principle of "Restore first, repair next", we have built a process to enable the technician to examine the incident for service impact. Next, to use the appropriate M&Ps to restore the service, and finally follow through with the M&Ps to repair the service.

For example, in the case of a simple over-head cable-cut due to a damaged pole, the technician could use the procedures for erecting a temporary pole and splicing the cable to restore the service, and then attempt to fix the root cause and move the service over to the permanent pole over a slightly longer but not customer-impacting time frame.

2.5 Tools for Transformation

2.5.1 T-Plan/Q-Plan

Transformation is a multi-year program. Through this program, the CSP must build a series of capabilities (e.g., content storage platform, content distribution network, IP/MPLS core, VPLS enabled Edge, VPLS enabled Aggregation, OSSs supporting new network Elements, workforce training) across all processes and the network.

Once these capabilities are developed and verified, new services (e.g., Video-on-demand services, Interactive video services (e.g., gaming), VPLS, Quote-to-cash processes and systems) can be built on top of them.

To program manage this massive undertaking, we have developed the BII practice for program management. It includes three major components: (a) T-Plan (Time-Plan) for Capabilities, (b) Q-Plan (Quarterly Plan) for committed capabilities, and (c) a Q-Plan for service introduction.

The T-Plan is an architecture view for the network capabilities over the next few years, up to 5 years. It is important to note that the T-Plan does not represent the funding commitment, it is purely a logical step-by-step plan to develop and deploy network/service capabilities. Also, this plan should be developed and owned by the CSPs' technology Labs.

T-Plan is a collection of capabilities that must be used by the CSPs' services overtime to fulfill the full scope of their target network and services. The grouping of such capabilities is based on (a) availability of the technology, (b) key business drivers for revenue generation, cost reduction, etc., (c) and the sequencing required for technology insertion.

The Q-Plan is a time-based view for the network capabilities outlined under the T-Plan. It is important to note that the Q-Plan does represent the funding commitment, and it is committed by all CSP's processes, as well as the vendors, impacted by the transformation. This plan must be developed and owned by the Program Management Office.

2.5.2 Zero-Based Workforce Model (BII-Zero-Based Workforce)

Zero-based workforce modeling (Ovstoll, 2019) and (Mastel, 2003) is aimed at planning for staffing levels for the CSPs' ETE processes, given a specific state of the network technology and automation that is in place, or being planned for. This modeling approach is also key in benefit realization associated with technology upgrades and/or future automation initiatives.

This model is built for each business process (e.g., Provisioning, capacity planning and deployment, network operations, etc.). Some business processes do not lend themselves to this level of modeling, e.g., research function. However, some 80–90% of the CSP workforce can be modeled with this approach.

This model must be built for each type of work drivers, work volume, including activity time within the process. Total process time is then multiplied by the volume of the work drivers to arrive at the headcount "Full-Time Equivalent" for the process. This method is key to measure the impact of transformation on a specific process and the associated headcount. The results will provide vital input for the development of business case (cost/befit analysis) for the transformation throughout the transition from the PMO, to the TMOs to the FMO.

2.5.3 ETE Service-level CAPEX Unit Cost/Platform Unit Cost

There are 100's of pieces of network elements that are pieced together to connect a customer to the contents/users. The CAPEX in the ETE connectivity must always be the target for massive cost reduction. The key and fundamental question is: what is the ETE CAPEX unit cost for generic services? This question must be followed by the next question: which component of the ETE service has the most impact on the ETE CAPEX unit cost?

You want to build the ETE CAPEX unit cost model for at least 3 major generic services that are provided through your legacy networks: (1) Mobility service (measured by equivalent Million minutes of usage—it is key to establish this metric based on the mobility content for video, data, etc.), (2) wireline Data service (for consumer/business customers, measured by equivalent GE), and wireline Video services (for consumers, measured by Living Unit Passed).

Then you need to build the reference connectivity flow, from the customer premise through the network to the on-net content and/or to the IXP exit point for the off-net content. The key is to use the unit cost for every component of the reference connection to establish what the true CAPEX unit cost is for your legacy networks and services.

This model is key to determine where the target for CAPEX unit cost reduction must be focused on first. It also identifies, with laser-sharp accuracy, which network elements the focus of unit cost reduction should be placed on, to have a material impact on the ETE CAPEX unit cost. Also, it ensures if the CSP has the right CAPEX improvement projects identified to achieve the transformation targets for the CAPEX unit cost reduction. This unit cost model must be built for both Platform Unit Cost (such as content platforms, Intercity transport, Common systems, etc.), as well as Service Unit Cost, with a focus on particular slice across multiple platforms (e.g., SoIP such as interactive video).

The ETE service CAPEX unit cost model is key for targeting specific transformation as it relates to the 80/20 rule, for which NEs to address in priority sequence.The Platform unit cost model is key to ensure that the NEPs are being focused on the YOY cost reduction for each component of the ETE connectivity and the services.

Once the FMO unified network is built and is in service, The same CAPEX unit cost model must be built for the FMO network, with the focus on the seven service flows, particularly the bandwidth (wireless, High-speed internet access (100's Mbps), and high speed data (Gbps) access), streaming, and interactive Video.

2.6 Service Disaster Recovery (BII-SDR)

CSPs must architect and build their network with the Co0 for demand-based maintenance, i.e., the architecture for the network and the services must be resilient under well-defined failure conditions without any impact to the customer services.

However, for any failures beyond the boundary conditions for resiliency, CSPs must be prepared to treat that as a catastrophic event, and must be able to respond in real-time with Service Disaster Recovery practices.

We recommend that you develop the concept of SDR to enable CSPs to recover from catastrophic failures within 72 hours (Wrobel and Wrobel, 2009). It requires 24/7 monitoring via global operations center, 3CP, real-time process to respond, trained personnel, plug-and-play platform architecture that lends itself to being placed in trailers waiting to be deployed, and the SDR data-based profile for all Data Centers, accessible by the SDR Command & Control for deployment.

The key to the success of the SDR is frequent test exercises, preferably every quarter in an operational setting.

2.7 Transforming CSP—Accountability, Goal Setting

Accountability and organizational transformation are critical factors for sustainability after the completion of the transformation.

Accountability must be redefined from "functional" to "ETE processes", supported by ETE KPIs. The decision-making is also transformed in two ways: (1) *day-to-day decision making* which must be decentralized and controlled by the working level process associates and guided by well-defined engineering rules and the KPIs, and (2) *decision making for structural changes* (such as network or systems architecture, Process, etc.) which must be orchestrated in a centralized fashion, to optimize for business performance.

Performance goal setting process must be transformed with the key objective to cascade DMOQs (Direct Measure of Quality) to work steps with a well-defined KPI target at each work step. This fundamental cultural change ensures that every process associate's daily actions are aligned with the business goals:

1. *Direct Measure of Quality (DMOQs)* is often aimed at competitive positioning targets which must be set by CEO/CFO and the senior executives, e.g., revenue target, cost reduction target, customer experience performance target, etc.
2. *Business Unit (BU) business goals* established, and cascaded from DMOQs, e.g., new services revenue goal, new services from footprint expansion goal, network/service performance goal.
3. *BU performance targets* established, consistent with delivery of BU Business Goals, e.g., market expansion goals, service provisioning cycle time reduction.
4. *BU ETE Process KPI targets* established to operationalize the BU Performance Target, e.g., customer Ordering cycle time, Provisioning cycle time, service activation cycle time.
5. *Work step KPI target* established consistent with the desired performance of the work steps in the ETE process.
6. *Specific initiatives* identified and funded to deliver the performance established instep 4.

A program management function is key in ensuring that all initiatives are on track for benefit realization, performance delivery, and proactive course-correction when needed.

2.8 Transforming CSPs—Setting Transformation Targets

CSP transformation impacts all aspects of the business, including customers and customer experience, network and services, processes, OSSs/BSSs, and the security. A critical factor for the success of the transformation is to be guided by a set of well-defined business targets. A successful transformation should produce FMO and TMOs that are built and deployed to deliver the specifics for the following key competitive business targets:

(a) Revenue (new services, and footprint expansion),
(b) QoE Performance,
(c) CAPEX unit cost reduction
(d) OPEXunit cost reduction/productivity
(e) Scalability (network, services, processes), and
(f) New services.

Once the transformation Business Targets are established, the CSPs must translate those business targets into "design objectives" for every part of the business, such as the Network, Services, OSSs/BSSs, Culture, organizational structure, etc. The binding of the Business targets and the "design objectives" is key in ensuring that the transformational changes throughout the business are aligned to support the business targets.

Revenue Targets:

Revenue targets must be established to address projected market share on the existing footprint, the expanded footprint in each market, and the revenue from new services. A minimum 3% incremental revenue target is expected YOY for a tier 1 CSP, and a much higher revenue target is expected for Tier 2 and below.

QoE Performance Targets:

Today, the CSPs are being benchmarked for network performance in the six network performance categories: call performance, network speed, text performance, data performance, overall performance, and network reliability.

Over time, the benchmark is rapidly changing to include the 7 Macro Service flows (MSF): browsing, streaming, Interactive Video, VoIP, access bandwidth (consumer & Biz), and service availability.

CSPs must translate QoEs to performance targets for the MSFs. It is of utmost importance that the network is architected and built to support the CSPs' intended MSFs. CSP with quad-play services (video, data, voice, wireless) must design their Unified Network for the most demanding MSFs' QoE, i.e., Interactive Video with High-speed access to support high-resolution content.

We need to establish what these MSFs are, the customer expectations for how they must perform, and the associated performance targets that the network must be designed for, and operated at (latency, jitter, packet loss, availability).

CAPEX Unit Cost Reduction Targets:

Target setting for CAPEX, and CAPEX unit cost reduction must be focused on the business need to maximize Free Cash Flow. Free Cash Flow is the EBITDA minus CAPEX.

CAPEX is well managed when the CAPEX required for the traffic growth is invested in the architecture and network technologies with the lowest CAPEX unit cost. CAPEX target setting must take into account the ETE Service CAPEX unit cost, YOY traffic growth, and YOY projected revenue growth,to establish the CAPEX investment targets for the (1) "technology replacement" and the (2) "traffic growth". The timing between these two is key when optimizing the free cash flow.

OPEX Unit Cost Reduction Targets:

Target setting for OPEX, and OPEX unit cost reduction must be focused on the business need to produce a competitive margin.

The drivers for margin are defects and cycle time. The key is to reduce the "Reserve"(also known as Contra Revenue) and the cost of SLA compliance. Must prepare to reduce billing DPM (Mastel, 2003) by a factor of 10+ (Tennant, 2001), at the same time, the cycle time for customer and vendor-facing processes must be reduced to nearzero (ala the self-servicing model), (Stalk and Hout, 1990), e.g., ordering, provisioning, supplier transactions, etc.

Scalability Targets:

Network scalability is another key aspect for target setting. Ascalable network maintains its architectural intent for expansion YOY, and unleashes the sales force to respond to the customer demand in almost real-time. The network scalability target is different for each domain of the network, and must be established for each domain respectably.

Given the overall traffic growth of 30% in the access and 50% in the core YOY, we must establish scalability for the Core to a 10+ years before the new technology justifies a renewal, for aggregation and access to a 5+ years, and the content to a 3–5 years. This means that today's' technology selection decisions must be capable of delivering on this level of scalability over time given the planned traffic growth YOY. Also, the scalability of OSSs/BSSs must be established for: legacy services @ 10X, new services @ 20% of incremental revenue YOY for 5 years, and cannibalizing services @ the same rate as the legacy service.

New Services Targets:

To meet the revenue target, a good portion of the incremental revenue must come from new services that the CSP has never offered before.

Given our experience in the introduction of new services, we have concluded that some 20–25% of the incremental revenue should come from new services. For example, if the incremental revenue target is 5% for the CSP, then some 1–1.25% of that revenue should come from new services, such as VoD subscription, News/

Magazines subscription, etc. This target must guide the new service creation process and any associated investment that will be required.

2.9 Transforming CSP—The Network/Process

Generally speaking, there are three categories of Process/OSSs/BSSs in a typical DSP: (1) Services over IP platform (SoIP), (2) Customer/partner-facing platform, and (3) Network facing platform. A platform in this context refers to software/hardware/networking/process infrastructure upon which the platform is functioning. However, as we are not promoting a specific hardware nor software or networking platform in this book, we refer to a platform by its functionality.

The overall architecture must be flexible and efficient to enable the fast introduction of new and emerging services. These services must be available to both business and consumer customers over any access technologies (e.g., wireless 4G+, WiFi, fiber, copper) while ensuring the same user experience, independent of the device or the access technologies being used. Also, it is critical to support seamless handover of live sessions between any pair of supported access technologies or devices.

To accomplish this, CSP must transform its overall Process/OSSs/BSSs architecture to a single, common, and shared infrastructure for all services and one service logic for all access technologies. This infrastructure must be based on standards and must support a common run time execution environment.

Our intent in this book is not to describe the Process/OSSs/BSSs architecture for a DSP. The key is to describe the overall functionality that must be put in place in order to accomplish the above architectural requirements.

2.10 Transforming CSP—Security

[This section is planned to be expanded in the next release of the book. The outline below is aimed to provide an introduction to what is planned to be included in the next release.]

An IP connection is all that is needed in today's interconnected environment, rather than a fixed circuit connecting point A to point B. This is presenting an earth-shattering approach to the traditional security approach. In that, ETE cybersecurity is now comprised of a security apparatus for the users, access, aggregation, core, content, and the interconnections. The focus must be on the defensive aspect in preparation for the cyber-attacks.

The cybersecurity can be engineered and built for a closed network, or an open internet network:

(1) Closed Network: Enterprise Intranet—not the subject of a future release of this book
(2) Closed Network: National Intranet
(3) Open Network

In each case, the following key aspects of cybersecurity must be addressed:

- User security:
    - o User password/security certificate
    - o Event logs
    - o Validated access against agreement
    - o Enforcement of CA & PKI
- Access security
    - o Access control (On-net users, and network interconnect using digital signature, and/or certificates)
    - o Encryption at the interconnect
    - o End-user access through VPN, SSL
- Aggregation/Core security
    - o through IDS and Anti-DDOS (CISA, 2019) attack
- Domestic Content security
    - o National Email, Search Engine, supervised content
    - o National CA, Root DNS
- Content purity
    - o ACL, DPI, Intelligent Filtering

For a closed network, e.g., intranet, the cybersecurity must address the security of the users (through CA & PKI, VPN, SSI), access (through PKI), aggregation, the core network (through IDS and Anti-DDOS attack), the content and the associated services (browsing, streaming, IV, VoIP, Email, Search Engine, fail-safe internet, supervised content) through DPI & filtering. It also must address the legitimacy of the external content and services through DPI, filtering, and intelligent filtering. Overall, must also be addressed through the Disaster Recovery apparatus.

For an open internet, it must address the same concerns as the closed network, except for the services that can be provided in the open internet, such as CA, Email, Search engine, external content, etc.

2.11 Transforming CSP—Operations Discipline (BII-OD)

Based on our extensive experience with planning and deployment of global tier 1 network, we often observed, before the transformation, that poor legacy network performance was being tolerated by the operations team, with excuses such as the pending introduction of promising new technologies and networks that promise to fix the shortcomings in the QoE. This mindset is flat wrong. Your legacy networks, no matter what the underlying technology, must always be optimized to perform to the BII QoE.

The Network Operations function must ensure a stable and high performing network given any state of the legacy network and services technologies. In that, we established the flawless execution of the day-to-day operation to be the most critical factor for success for the transformation as well as the on-going customers' support.

We have identified the following BII Operations Discipline to be critical factors for success for flawless execution and operations:

(a) A working-level operations dashboard, well-integrated with the QoE Dashboard...to enable proactive/preventive/predictive decision making, to keep the network operating at peak performance.

(b) Customer communication...to ensure that customers are proactively informed of their network performance.

(c) Ask Yourself (AYS)...to ensure whoever is touching the live network is well trained and equipped with the procedures, oversight, and all the tools needed to carry out the operations activities flawlessly.

(d) Network Event...to ensure that a planned network-touching-event is performed without any impact on the customers

(e) 3CP/TCB/MCB...to ensure that the right decisions are made through a well defined command-and-control process to recover from a customer-impacting incident

(f) Root Access...to ensure no unauthorized access (e.g., Labs personnel) causing unexpected and unintended network incidents is allowed

(g) Outage Management...to ensure the outage is managed through process management, and to ensure that the root cause is being institutionalized, near real-time, in the business processes

(h) M&Ps...to provide a tool for all process associates, regardless of their years of experience, to deliver the same expected performance for the care of the network.

(i) Disaster Recovery...to ensure that when all fails due to catastrophic events, the network is restorable to its original performance level within a pre-specified time.

This list of Operations disciplines, once implemented with the right processes, systems, and training, will ensure that the PMO is operating at the design objective level for QoE.

2.12 Transforming CSP—Transformation Culture (BII-TC)

E2E accountability and E2E Process ownership is a key to cultural transformation. Even if a process associate is responsible for a small function in the ETE process, he or she must understand that the success is defined by how well the ETE process, that he or she is engaged in, is performing. This fundamental cultural transformation is crystallized by the Golden thread from the business goals, to E2E process goals, to organization goals, and cascade to individual goals.

We need to create a culture for risk-taking while being held accountable. Employees must learn to understand the 90% of the bottom of the iceberg from the 10% on the top. This requires the employees to be less risk-averse, have more sense of urgency to attend to the issues, their behavior must be driven by the KPIs, everyone must understand that they are part of an ETE process, and their sole existence in the business is to deliver measurable results for the ETE process.

The decision making, for the process associates, must be decentralized for day-to-day operations (there must be no bureaucratical processes, including approval-to-spend), and yet centralized for structural changes. The process associate must be empowered to make decisions, including financial, all driven by well-established rules. This approach must be carefully supported by a real-time dashboard that cascades customer-facing performance to the root causes throughout the network and the processes.

2.13 Transforming CSP—Investment Monetization

One of the unique difficulties of transformation on the scale we attempted, was to prioritize the dozens of initiatives that were already in the process, or were being created during the transformation.

To handle this challenge, we employed a decision-making matrix—called the discounted pay period (DPP) (Kenton, 2019). This matrix was critical in quantifying the commitment to individual initiatives and applying resources according to the Schedule-of-Authorization. The process involved careful thinking about how much time any single project needed in order to return an investment—and how much approval it would require for investment decisions. The streamlined approval process must prioritize in-year saving vs. near term vs. strategic investment decisions.

This concept is a cooperative effort, and must be led between the President of the network and the company's CFO. It works like this: in general, any initiative with less than one year projected to full ROI could get funded with the president's approval rather quickly. Actions with greater than one year and less than three years DPP could be approved by the CFO and the president if the two were in agreement. If the initiative had a DPP longer than three years, it was considered strategic and required board/CxOs approval for any investment—think massive undertakings like fiber deployment through consumer footprint or new wireless technologies/spectrum acquisition.

In general, 80 percent of new initiatives had less than a year discounted pay period, 15 percent had a DPP between 1–3 years and 5 percent fell into the "greater than three years" category. By stratifying initiatives by DPP, we were able to quickly pour energy and capital into programs that would inject the company with revenue and cost savings right away. More complicated and ambitious projects would always get the consideration and debate they required. However, in general, the discounted pay period meant that the company would refocus its attention on the most beneficial, low-risk initiatives first.

2.14 Transforming CSP—Business Case for Transformation

The content and format of a typical Business Case for telecommunication project is not within the scope of this book. However, we are addressing the key aspects of the models that must be developed and maintained to support a business case, and to hold organizations accountable for the benefit realization of the transformation program.

The business case for transformation must be supported by the following:

(1) Zero-based Work-Force model (Ovstoll, 2019) for each business process. This model is critical to establish the benefit realization for transformational investment.

(2) ETE CAPEX Unit cost model at the service level. This model is the driver for laser-sharp focus on transformational utilization of the CAPEX expenditure.

(3) Revenue model, to ensure the introduction of process improvement and new services do follow the rule of 72, in that the new incremental revenue and/or cost-saving could be triggered in January through September of each year.

(4) Cost Avoidance model (NASPO Benchmarking Workgroup, 2007). This model must provide an accurate presentation of cost avoidance for the specific aspect of the transformation.

(5) Customer retention model (Editors: Huber et al., 2008). This model must present a fact-based financial presentation of the impact of customer flight given a specific status of the PMO and or TMOs.

2.15 Transforming CSP—Transformation Team

The picture below (Figure 2.7) outlines our collective experience for the most effective structure for CSPs' transformation. The scope of the team includes planning, development and implementation, process management, dashboard, optimization, and benefit realization. The overall program is supported by a program management function.

We have identified a critical function, named "Forcing function" which must be staffed by the key executives holding functional responsibility across CSP. This function is to resolve cross operational issues and to enforce the execution of the transformation plan.

Another key function is "Strategic Advisors" with the sole responsibility to provide an independent technical and operational direction.

It is important to note that the CSP's select team, as depicted in this structure, will be intensely involved in helping and leading with the transformation.

It is of great importance that the external entities, such as business customers, interconnecting CSPs, and regulatory agencies are connected with the transformation effort throughout the full planning and deployment process.

2.16 Transforming CSP—High-level Steps & Time Line

Your current legacy networks and services must always function to deliver BII QoE, at a predictable cost for the SLA compliance. As a result,a prerequisite to transformation is the focus on the legacy networks for network & service optimization.

Assuming that the legacy network is already optimized and is delivering BII QoE, the CSP is ready to undertake transformation, and it is a three-step process, as

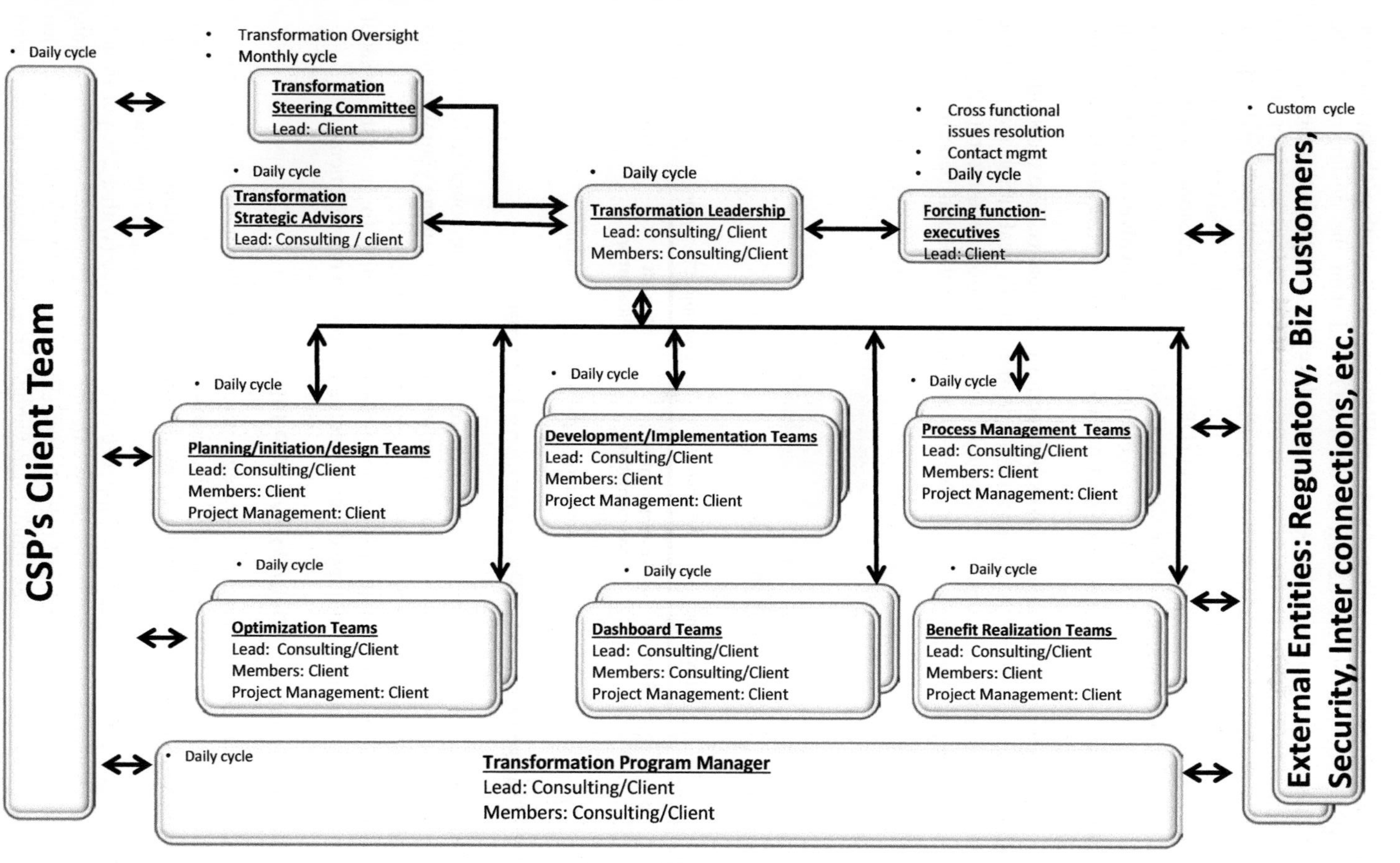

Figure 2.7. Transformation team structure.

depicted in Figure 2.8: (1) Transformation Planning: starts with a characterization of PMO, followed by the design of the FMO, and then the design of TMOs; (2) Transformation Implementation: starts with development and deployment per the T-Plans and Q-Plans for TMOs, migration from PMO to TMOs, and keeping PMO performing at near 6 sigma throughout the journey to FMO, and (3) Transformation Benefit Realization: supported by program management, business cases, and driven by the concept of Discounted Pay Period (DPP). DPP is a decision-making matrix. This matrix is a critical factor for success, used to quantify CSPs' commitment to certain initiatives and apply resources accordingly. This involves careful thinking about how much time any single project needed to return an investment—and how much approval it would require.

The transformation planning takes between 3 to 6 months. The transformation Implementation is a 1 to 3-year program, and the benefit realization starts at the planning stage and will continue through implementation, ending roughly one year after the transformation is complete, to account for the total benefits and the deliverables.

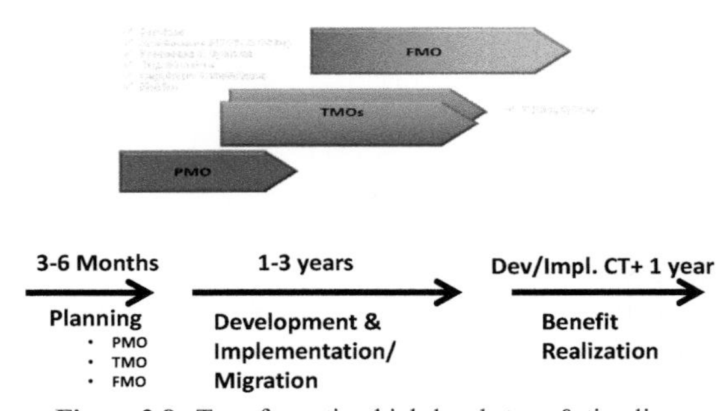

Figure 2.8. Transformation high-level steps & timeline.

2.17 Transforming CSPs—Lessons Learned

When you are embarking on a massive transformation, you are bound to learn many lessons as you push through unchartered territory while transforming your workforce, the culture, the state-of-the-art technologies, and the BII practices.

Here, we summarize some of the top-of-the-mind learning for us in the areas of investment decision, technology development, and process management.

Strategic Investment Decision vs NVP Pitfalls

Telecom transformation requires a strategic investment decision, which is contrary to traditional justification through NPV (Hopkinson, 2017). CSPs' cash cow has been constantly under attack over the last few decades and will continue to experience the same going forward. CSPs, at times, must call for strategic investment to stay relevant going forward. See what happened to the LD voice (with a 25c per minute revenue stream in the 90s), replaced by wireline VoIP at a fraction of the revenue, soon after replaced by wireless VoIP at near-zero revenue. At the same time, video streaming became the center stage for revenue generation demanding broadband connectivity, specifically at the last-mile. Many CSPs were caught unprepared for this revolution. Even those who took the action to go with fiber optic in the last-mile, were forced to rethink their approach with the force of the NPV.

One of the key lessons for us was that the telecom transformation requires strategic investment decisions, from time to time, which is contrary to traditional justification through NPV and ROIC.

Technology Development/Deployment Pitfalls

- Co1 is meant to be applied to all aspects of the business. It is a powerful and effective concept. However, in some cases, to achieve the time-to-market objective, it is wise to apply Co1 in stages, e.g., the build-out of the Co1 OSSs/BSSs infrastructure as-a-first-step may take longer cycle time with higher CAPEX requirements. A better Co1 staging approach is first to build a Co1 customer-centric platform/bus DBOR (data warehouse-Database of Record) to fix the data accuracy from sales-to-billing to near 100% accuracy level, with a focus on a single customer view of all services in use. With DBOR, build interfaces from all legacy OSSs/BSSs to the DBOR to establish a single customer view under strict data dictionary management for sanctioned sources of data elements—this can be achieved fairly quickly, then over time consolidate OSSs/BSSs to achieve Co1 for each group of the legacy systems.
- Don't transform with custom-built technology: at times, we were pushing bleeding edge technologies into the network. It is of utmost importance to not custom build any components of the overall network. In the late '90s we were pushing for custom Next Generation Network switches that we decided to build to accommodate for our scale and projected growth…big mistake! Our learning was that it is not the technology that differentiates a big-league DSP from a "me-too", but it is the DSP's architecture leveraging the off-the-shelf technology that makes the difference.
- Don't bet on start-ups and small vendors for strategic elements of the network. It makes the CSP hostage to a single-source supplier, and all the associated pitfalls for scalability with a start-up. When the only choice is to go with a small/startup vendor, make sure a major vendor/partner provides the supervision and accountability for the commitments from the start-up. Corporations, pragmatic partnering, smart integration strategies and/or ultimately acquisitions of technologies from start-up companies can provide a significant competitive advantage in the market place as we have experienced in the long haul DWDM optical systems market in the early 2000s. Technologies at that time provided 10 Gbps on each of the 80 wavelengths. However, CSPs were facing exponential growth in their Networks and looking for more cost-efficient methods to grow their network capacity. At that time, 40 Gbps was the cutting edge of optical technology. Many R&D programs were underway, but no 40 Gbps products were yet available. The incumbent supplier could offer 40 Gbps as a new feature under the terms of the existing supply agreement

but the lead time for 40 Gbps product development put the projected product availability well beyond the date required to cope with market demand and to at least maintain the CSP position as the long-haul DWDM supplier. To minimize development efforts and get the product to the CSP as early as possible, a pragmatic partnering and integration strategy was adopted from a small start-up company in Silicon Valley under the overall leadership and integration of the incumbent supplier. Their hardware and software would exist in physically separate equipment bays and integrated with the incumbent suppliers' system only in the management system software. Other than that, there was no integration of hardware and software with the start-up's hardware and software. From a network management and operational point of view, the combination looked like one system, but from a development point of view, integration efforts were limited to the management system software. This approach allowed the incumbent supplier to meet the CSPs time-to-market requirements and maintain a sole supply position in the long-haul portion of the CSPs network.

The lesson was that sometimes key strategic elements of a network require an intelligent, pragmatic and smart integration approach together with leading-edge start-up companies to meet time-to-market requirements and maintain a competitive advantage.

- Don't let patchwork influence your investment decisions: for example, short-term cost avoidance decisions for fiber deployment may force the CSP to bury fiber vs. building through a conduit. These actions could come back to haunt with significant fiber impairment over time, and could make the fiber replacement very difficult and costly.
- Don't focus on forecasting for the right-capacity in your DSP network, but focus on making your network scalable in near real-time. We learned, time and time again, that the forecasts for network expansion were hard to predict and were often wrong. You will be caught with surprises when you cannot respond to the customer demand given the long cycle time for network expansion and the build-out. We learned that the network architecture had to lend itself to real-time scalability for the on-net customers, and must be prepared for near real-time expansion to serve the off-net customers.

Process Management Pitfalls

- There must be an arm-and-a-leg separation between the Process Owners and the owners of the operations. The conflict of interest is detrimental to the transformation. Don't get the process owners to report to the operations management: at times we had the Process Owners reporting to the Operations Managers. This conflict of interest slowed down the transformation process. For example, the integrity of the zero-based workforce model could be compromised significantly and slow down the transformation.

CHAPTER 3

Assessment of the Present Mode of Operation (PMO)

3.1 PMO: Introduction

Fact-based assessment of the PMO (Harmon, 2002) and (Introduction to eTOM White Paper, 2017) is the first and a major step for the transformation, it is also essential for establishing a meaningful path forward. It helps to build a strong business case (Lema, 2017) for the transformation, leading to the plans for FMO and the TMOs, and the development, deployment, migration, and benefit realization.

The assessment must provide you with a fact-based understanding of how your network and services are doing concerning the performance of QoE (Chen, 2012) for both the QoE for the digital services, as well as the QoE for the touchpoint services. It should also address how well it is performing competitively on customer-facing service performance, how efficient is the CAPEX (Telecom Capital Expenditure, 2020) and OPEX (Taga et al., 2020) unit cost contribution to the Margins and the Free Cash flow (Jacobs, 2020). It should also clarify if the CSP is on track for introduction of new high margin services and revenue growth going forward, and what regulatory conditions must be negotiated to enable fair competition as the CSP is expanding domestically and globally.

This analysis should provide you with all you need to build the business case for the transformation, to prepare for a company-wide communication for the "why" for the transformation, and to provide the customer communication for what they should be expecting throughout the transformation.

Here is what we have done time after time and recommend to use for the assessment of the PMO. This assessment is to be completed to develop a snapshot for the CSP through two different filters:

(1) First filter, through the formulation of a comprehensive questionnaire, and subsequently working with the key organizational process owners to secure the fact-based answers to the state of the CSP. The outcome should solidify the network and OSSs/BSSs architecture, ETE services, service operations, service performance, and the key issues with solid/fact-based quantification for any competitive shortcoming of the network.

In the context of this book, we will not go into the details regarding process management. In any CSPs where the process management discipline is not fully deployed and practiced, the characterization of the PMO, for the first time, will prove time-consuming and rather difficult. In this book, we are not attempting to describe what must be done to establish and deploy process management discipline, which we believe is a must for all CSPs to have implemented. Instead, we describe the key issues that must be researched to establish a fact-based characterization and the associated shortcomings of the critical components of the PMO.

(2) Second filter, through network and business data mining and analytics. We strongly recommend implementation and comprehensive analysis of the *business probes*, as well as ETE real-time *network/services probes* (Chu, 2014) to provide a full assessment of the business. The construct of this second filter is unique and the associated metrics must be built by the assessment team, based on the principles of QoE and the analysis of the business vital signs.

Overall assessment of the PMO is critical to establish the case for the transformation. Through the first filter, we learn about the plans, the practices, the processes, and through the second filter (data mining and analytics) it becomes crystal clear as to the effectiveness of the plans, and how the plans have been executed, as well as the specifics for how the network operations are performing daily. It is through these two filters that we learn all we need to know about the CSP from the angle of the customers' experience.

The size of the gaps between the findings from these two pictures must raise the alarm. The gap between the BII-practices, as described above, and the PMO practices is even more alarming. Alarming as to whether the customers' expectations are being met, whether the plans and practices for the network/services and the operations were solid, whether the execution of the plans was carried out flawlessly, and whether the network operations is doing its job in ensuring that plans are not compromised over time by the day-to-day changes that are being made in the network; and we can assure you, on this last point, that the CSPs' day-to-day operations continuously compromise the integrity of the overall architecture. The key is to ensure any compromises are automatically audited and made visible through a real-time dashboard.

At the high level, we have learned all we needed to know from the above two snapshots of the business to build the business cases leading to many successful transformations for many of the CSPs.

A successful business case must address one or more of the following issues:

(A) **Low margin service portfolio:** limited service portfolio primarily focused on connectivity services vs expansion to include digital services.... Look for *up to 25% improvement in the EBITDA margins.*

(B) **Not competitive QoE Performance:** poor QoE...it is a *kiss of death for losing market share, and increased and costly churn rate.*

(C) **High CAPEX unit cost (Telecom Capital Expenditure, 2020), associated with operating multiple networks:** multiple redundant networks for IPs,

TDMs, Data/s, voice/s.... anticipate *up to 80% CAPEX unit cost reduction by consolidating into one, scalable, unified Network.*

(D) **The shortcoming associated with the ETE processes that project a not-easy-to-do-business with:** long cycle time in customer-facing processes, e.g., ordering, provisioning, restoration, as well as poor coverage....*churn rate of up to 75% improvement...estimated equivalent up to 2% of revenue, reduction in OPEX.*

(E) **Inefficient OPEX unit cost structure (Taga et al., 2020):** expect *up to 80% reduction in the OPEX unit cost.*

(F) **The high cost of "reserve":** significant size of the "Contra Revenue" is what to attack...this is a low hanging fruit, and it produces *up to 10% of the revenue in improving free cash flow and the margins.*

(G) **High cost of SLA compliance:** expect *up to 5% of the revenue, improvement in free cash flow and the margins.*

We will cover each one of these issues in this book.

3.2 PMO: ETE Services to be Assessed

The breadth of CSPs' services is endless. It covers a range of connectivity services for both consumers and business (primarily bandwidth), as well as content services,such as interactive video, video surveillance, cognitive personal assistant/financial planner, IPTV, 4K+OTT, IoT, privacy services, cognitive SMEs, ambient intelligent solutions, and an incarnation of these services.

The key question is what should we be looking for in the assessment of an endless number of CSPs' services? Is it the video services, home automation services, home monitoring services, variety of VPN services, or MIS, or HSIA? Or P2P connectivity services, or wireless services in all operating markets? Yes, it is a daunting task if you approach assessing the ETE networking and service platforms in this way.

Our collective experiences have shown us that all current and future CSPs' services are primarily provided through 7 major service flows (Figure 3.1): (1) browsing, (2) streaming, (3) VoIP, (4) Interactive video (IV), (5) service availability, as well as 6&7,) appropriate access bandwidth for consumers and business customers. If we perform data mining and analytics of the performance of these 7 macro service flows, we are done with a complete assessment of all services being offered by the CSPs.

The key driver for the last mile band width is driven by the bitrate for the content. Let's keep in mind that the bitrate of the content, for consumers, is continuously changing from SD, to HD, to 4K+, and the future bitrates for true 3D and holographic contents will continue to challenge the access bandwidth for consumers. The bitrate is primarily impacting the last mile connectivity, and the scalability of the network and the content platform. The last mile connectivity is driving the push for future high-speed, symmetric last-mile access technologies through wireless connectivity such as 5G+, and/or wireline technologies for fiber-based connectivity.

It is important to expand the scope of the services, to also include the effectiveness of customer-facing business processes; we will describe the details about the performance requirement for these service flows later.

Figure 3.1 depicts the outline of the Macro Services Flows:

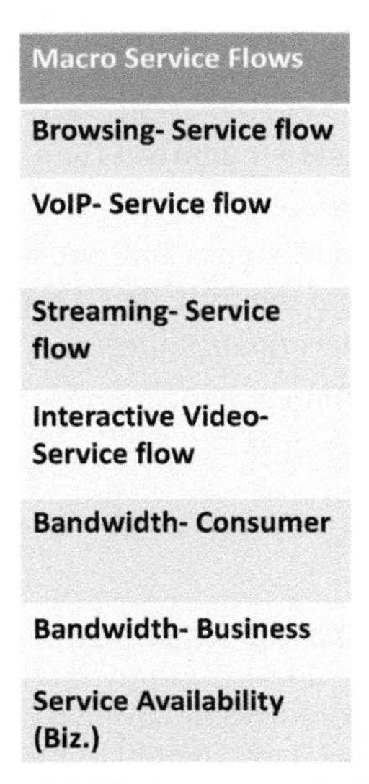

Figure 3.1. The 7 macro service flows.

We have assessed many CSPs, and here we are going to share some of the learnings from our past experiences. We are breaking it down into the performance of their services/networks, their overall OSSs/BSSs effectiveness, CAPEX efficiency, and OPEX efficiency.

3.3 PMO: ETE Service/Network Performance

Traditional CSPs offer mainly connectivity services (e.g., internet), and in some cases, content services (e.g., VOD).

The first thing we look for in the assessment of the PMO is the Closed-Loop Performance management system. The following picture (Figure 3.2) presents a high-level view of the closed-loop system with 6 major components to assess:

We look through the CSPs' dashboards for weaknesses in the closed-loop-operations of the CSPs. We look for weaknesses in how the network/services measurements, QoE, and alarms are collected and calculated (1); how the alarms/ sensory data are converted to real-time dashboards (2); if data collection is based on peak traffic- 95 percentile?; if there is a drill-down capability to aid with the root cause and the surveillance, analysis, and the notification of the operating units (3); if the analysis is focused on preventive/proactive/predictive events?; how the operating units translate the notifications into actions (4, 5, 6); and how the actions are executed in a timely and proactive manner.

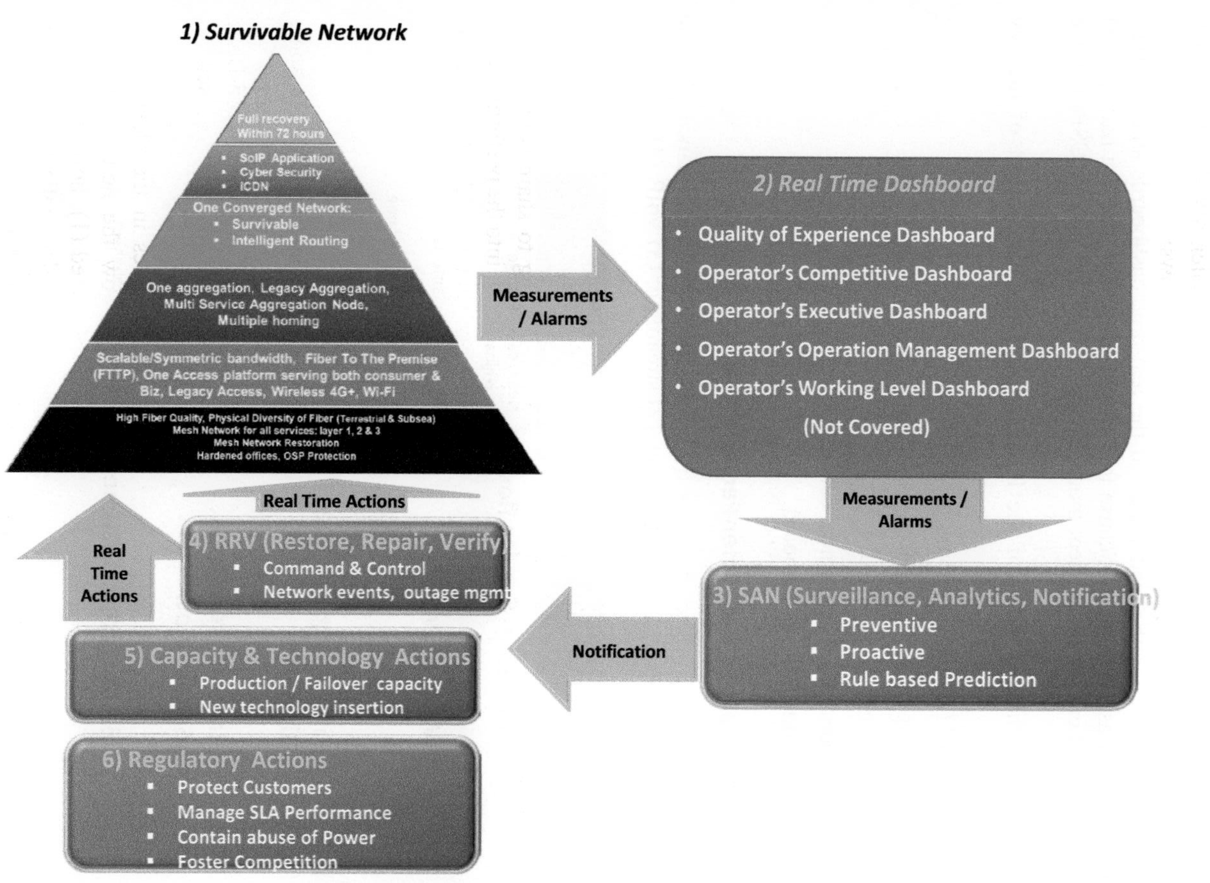

Figure 3.2. CSPs' closed loop performance management system.

In all cases, we found the performance of the services inadequate from a QoE perspective, e.g., poor interactive video experience, dropped calls, poor quality of voice, poor coverage, video pixilation and freezing, low availability of services, long cycle time to provide business services, etc.

We have correlated the inadequacy of the real-time dashboards and the drill-down capability as the main root cause for poor performance, and that had resulted in a reactive and crisis management mode of operation in most of the CSPs that we have assessed.

Let's walk through some of the learnings from the assessment of the key dashboards:

I. CSP's QoE Dashboard (Bouraqia, 2019)
II. CSP's Competitive dashboard (RootMetrics, 2020)
III. CSP's Executive dashboard
IV. CSP's Operations management dashboard
V. CSP's Working level dashboard

3.3.1 Assessment of Customer-Facing (QoE) Dashboards

We used the following service flows for the assessment of the performance of all the legacy networks (Figure 3.3): (1) browsing, (2) streaming, (3) VoIP, (4) Interactive

Macro Service Flows	QoE Expectation
Browsing- Service flow	• Near real time page download
VoIP- Service flow	• Voice quality Same as wireline • Call set up cycle time same as wireline
Streaming- Service flow	• Zero pixilation • Zero freeze • Sync'd voice & video
Interactive Video- Service flow	• Same as streaming • Action Trigger sync'd with video
Bandwidth- Consumer	• Guaranteed throughput (DL/UL)
Bandwidth- Business	• Guaranteed throughput (DL/UL)
Service Availability (Biz.)	• SLA compliant

Figure 3.3. QoE for the 7 macro service flows.

video, (5) Service Availability, as well as (6&7) appropriate access bandwidth (wireless, WiFi, wired) for consumers and business customers.

We built the QoE dashboard to assess the customer experience on the above 7 service flows, as well as the drill-down capability to establish the root cause for such performance throughout the network footprint, including access, aggregation, backbone, interconnection gateways, and access to the contents both on-net and off-net.

This QoE dashboard can be and must be developed and deployed quickly (in a matter of weeks) if it is not in place already.

Based on our observation, this QoE dashboard was often not in place, or if it was in place, it was to support a subset of the business customers, mainly managed services for enterprise customers. Here (Figure 3.3) is what we have established for the expectation of customer experience for these key service flows:

Next, we converted the above customer expectations into specific service level performance at different bit rate/FPS (e.g., latency, jitter, packet loss, bandwidth, availability) through extensive lab trials, customer feedback process, as well as live customer feedback. Let us take you through examples of some of the learnings from the past CSPs' assessment.

3.3.1.1 Overall Assessment of Real-time QoE

We would like to point out that the data points provided below are not from anyone particular CSP (Bouraqia, 2019). It is based on our collective experiences from many global and national CSPs.

Here is (in Figure 3.4) what we used as a benchmark for the performance of the 7 service flows provided by the CSPs:

Based on our analysis, most of the CSPs did not have a real-time QoE performance monitoring dashboard in place for the key Macro Service Flows. However, most of the CSPs did have realms of data on jitter, latency, packet loss, and bandwidth for different components of the CSP network, but few could identify a customer-centric ETE real-time measurement, where the QoE was being measured and monitored throughout the footprint of the CSPs' network, and during the peak traffic time.

In most cases, we had to build a real-time QoE performance data mining and monitoring dashboard in preparation for the assessment. For assessment, this dashboard is fairly easy to develop and to deploy quickly in a matter of weeks.

Based on our experience, and before the transformation, we examined the CSPs' own provided network performance data, which raised more questions than answers. However, given the performance that we gathered through the QoE dashboards, that we built for the assessment for each of their networks, and given the customer-centric threshold for acceptable performance during the peak hours, we learned a different story about the CSPs' QoE, and the ETE network and service performance.

The framework for the Real-time QoE dashboard that we deployed was based on a series of QoE probes: "Initiators", "Responders", and "Initiators/Responders". These probes were established on the network elements and/or on dedicated probe devices throughout the ETE network from the customer premise to the contents, including the access, regional aggregation, core, internet gateways, CDNs, as well as

Macro Service Flows	QoE Expectation	Acceptable Service Performance	Unacceptable
Browsing- Service flow	• Near real time page download	Page Download: < 3 Sec	Page Download: > 8 Sec
VoIP- Service flow	• Voice quality Same as wireline • Call set up cycle time same as wireline	Latency: < 170 ms Jitter: < 10 ms Packet Loss < 0.01%	Latency: > 250 ms Jitter: > 30 ms Packet Loss > 0.04%
Streaming- Service flow	• Zero pixilation • Zero freeze • Sync'd voice & video	Latency < 100 ms Jitter < 10 ms Packet Loss < 0.01%	Latency: > 120 ms Jitter: > 30 ms Packet Loss > 0.04%
Interactive Video- Service flow	• Same as streaming • Action Trigger sync'd with video	Latency < 80 Jitter < 10 ms Packet Loss < 0.01%	Latency: > 100 ms Jitter: > 30 ms Packet Loss > 0.04%
Bandwidth- Consumer	• Guaranteed throughput (DL/UL)	Guaranteed: Service flow bitrate /by market Symmetric speed on demand Bursting: 1G+ bps	No Guaranteed bandwidth
Bandwidth- Business	• Guaranteed throughput (DL/UL)	Guaranteed: symmetric, up to 40Gbps	< Biz demand
Service Availability (Biz.)	• SLA compliant • High availability	> 99.999 %	< 99.99 %

Figure 3.4. QoE performance benchmark during peak traffic hours.

the select off-net data centers. To ensure that the bottleneck was not the processing capability of the target websites/content provider servers, we also implemented "Responders" and "Initiators" at select co-located data centers where possible.

The picture below (Figure 3.5) presents the data mining architecture that we introduced for the QoE assessment. According to this architecture, we introduced thousands of probes on the ETE network. We set the data polling cycle at 5 minutes interval. The ultimate key point is to compile the performance data, based on 95 percentiles. This approach ensures that the network performance is properly measured during the peak hour, and maintained under the failures, consistent with the design objectives.

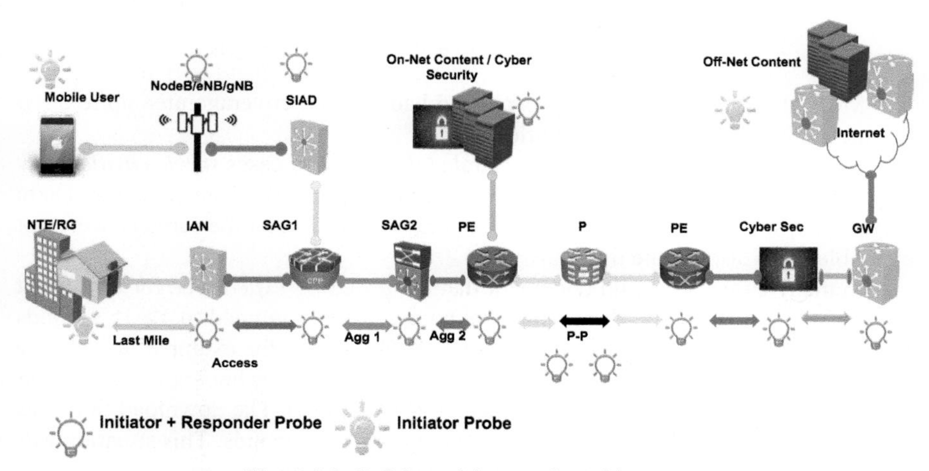

Figure 3.5. QoE data-mining—probe architecture.

For each of the 7-service flows, we employed minimum 4 key probes (HTTP Ops, UDP echo, ICMP echo, and UDP jitter Ops) to measure response time, latency (Almes, 1998), jitter (Demichelis, 1998), packet loss, and network availability for QoE, as well as for every domain of the ETE network as depicted in Figure 3.6.

Let's start with the breakdown of the traffic flow for the wireline and wireless 3/4G networks for the many networks that we assessed:

- **Browsing service flow:** ~ 30% of the traffic (web applications such as Google, FB, Adult content, CNN, weather, banks, real estate, etc.), followed by,
- **Streaming service flow:** 50% of the traffic… (live video such as Netflix, YouTube with ~ 20 seconds download/buffering, and live streaming such as Netflix and Pandora music with the quality of video varying depending on the bandwidth), next,
- **Interactive Video service flow:** 10% (gaming applications, e.g., Arcade, Action, Puzzle, Racing), and the balance of traffic volume spread across numerous services such as *VoIP*, social networking, pictures, Appstore, etc.

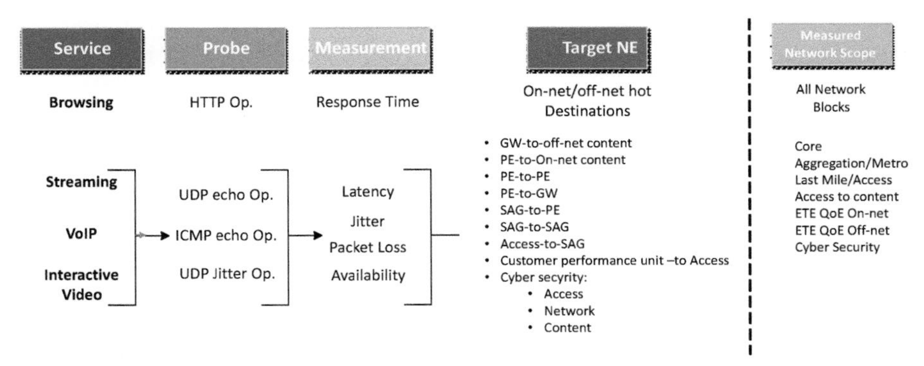

Figure 3.6. QoE data-mining—probe architecture & targets.

Most of the networks that we evaluated had an ETE coverage area up to 5000 kilometers in each region of the world.

For example, QoE analysis demonstrated that in many cases *service availability* (e.g., VOD streaming content, email) violated the benchmark as stated above. There were CSPs, suffering up to 300,000 DPM (i.e., one or more of the services were not available up to 30% of the time).

In many cases, we could document that the *browsing experience* for the on-net content was ok. However, during the peakhours, it easily jumped to >> 15 seconds for a download of a typical web page of 6–9 Megabytes,for example, a web site such as Amazon (primarily driven by the contribution of technologies such as JS [Java Script], CSS [Cascading Style Sheets], HTML, etc.). The download time was much worse for the off-net contents, in the range of 20+ seconds. This situation was confirmed for wireline and was worse for wireless customer experience.

In the case of *video streaming services*, during the peak hour, there were many cases of frequent buffering delays, pixilation, and lack of synchronization between video and audio, violating the benchmark.

In the case of *Interactive Video services*, e.g., the users of First-Person-Shooter (FPS) games, were experiencing unsatisfactory shooting performance. As the root cause indicated, poor performance is triggered at < 20 fps.

In the case of *Wireless voice/Text*,during peak hours, in several markets connected through the core, the call set up time was more than 6 seconds, and that was the threshold for the users to hang up and try again. The Accessibility (voice blocks) had a DPM of > 10,000, i.e., > 1% of the calls failed on the first try and had to be redialed. Also, systematically, we experienced more than 5 drop-calls (voice drops) in a stretch of 10 miles drive in many operating markets.

We also measured the delivery of the "*voice mail waiting indicator*" which was more than 20 seconds in many of the wireless provider markets, violating the competitive benchmark.

Also, we looked at the *customer churn rate* which was standing at >> 1.5% for many of the CSPs. When we looked at the reasons for the churn, some > 40% of the churns were attributed to the wireless/wireline network, and the rest to "competitive lure", "customer service", and others. Then we looked at the root cause of the network churns, the top causers were "problems initiating calls from home", "weak signals", "dropped calls", "poor voice quality", "voice mail notification delays", etc.

In the case of *Wireless Data,* we measured ETE latency over 190 ms on the 3G/4G network, in a number of the CSPs' markets, which impacted the Browsing/Streaming/IV services. We also measured the DL/UL throughputs, which were significantly under the threshold established for QoE on the service flows during the peak hour in key markets. Often, we learned that households with poor cellular connectivity, when enabling their smartphone with WiFi calling, experienced significantly more dropped calls as the uplink bandwidth of bursting to ~ 10 Mbps was insufficient to hold a VoIP, let alone a video connection, when competing with all the other data activities in the house hold.

The majority of the network churns for the Data network were due to: "difficulty connecting to internet", "problem maintain internet connection", "slow internet browsing speed", etc.

Then we watched over few weeks, the impact on the customer experience, over normal network failures (such as POEs [Plant Operating Error], fiber cuts, etc.). We also attempted to fail the select components of the network one at a time throughout the network footprint in order to assess the resiliency of the ETE networks.

For example, when we failed a network element, it was flagged as failed on the working level dashboard, buried among hundreds of other outages, however, it was not visible on the executive dashboards or could not show the impact on the QoE until the customer would call in.

For example,when we failed one fiber span between two core routers, the ETE latency went from 80 ms up to 380 ms. As a result, the VoIP application proved useless until the restoration was complete.

When we failed an edge router, it took more than 260 ms to recover from the failure. This long recovery time forced business customers (mostly financial institutions) to have an unacceptable delay in their trading transactions.

When we failed the connectivity to a data center for an on-net service, such as hosted mail&VOD service, the regional service was not available during the failure event. In a few cases, we failed regional fiber spans and had business customers experiencing service outages, even though the business customers had purchased redundant connectivity from their premise to the CSPs' POPs.

In a nutshell,the QoE analysis is extensive and reveals significant insight into the network architecture and deployment, and the network operations. We highly recommend to develop this dashboard for deep analysis and issue resolution, and to retain it on an on-going basis.

Here is the highlight of what we often came across from our assessment of the CSPs network through QoE probe investigations:

Customers:

i. We obtained fact-based understanding that the QoE performance was systematically poor during the peak hour, regardless of the CSPs' reported KPIs.

ii. Even though the CSPs were collecting data on latency, jitter, packet loss, and bandwidth, there was a lack of focus and ownership for the performance of

ETE service flows and QoE, as a result, customer-impacting failures were detected after the fact.

iii. SLA/QoE data gathering (through edge and customer premise monitoring) was not systematically in place.

iv. Poor browsing/streaming/IV due to long reach content.

Network:

i. Congested network

ii. QoE targets were not established, nor used to drive the network design to ensure resiliency for the key services; also, we quantified a high cost of SLA compliance.

iii. Network design not driven by QoE for the service flows: Browsing, VoIP, Streaming, and Interactive Video.

iv. CSPs' network performance data was often based on the average of the day-data and not on the peak hours when the customers typically have issues.

v. The connectivity architecture had many single-point of failures throughout the access/aggregation/content platforms.

vi. The ETE network was not built to sustain failures- Non-stable Core, Aggregation, and content under failure.

vii. Network designed in silos, with minimal focus on QoE

    i. Lack of CoS

    ii. Lack of ETE SLA management

viii. Slow access bandwidth, due to an abundance of obsolete access technologies, such as 3G/4G wireless and xDSL.

Operations:

i. Lack of process management practices that caused the learning from outages and anomalies to not be institutionalized in the process in real-time.

ii. Poor overall network and service performance due to lack of adherence to the BII Operations Discipline.

iii. In case of diverse fiber access connectivity, the backup fibers were often collapsed, and the few cases with similar experiences presented the indication that it was systemic, i.e., both primary and backup paths were often provisioned on the same fiber span. Additionally, we believe that the overall intention probably had been right for the diversity of the fiber paths, however, through subsequent network failures and restoration, the diverse paths had been compromised and collapsed, and were never repaired. Also,there were no automated audit processes to single out such cases for remediation.

iv. Inadequacy in service/network management practices due to alack of a Command/Control/Communication process.

OSSs/BSSs:

i. The significant volume of "contra Revenue" which was reflective of process defects and long cycle time.
ii. Lack of ETE e-Bonding for trouble resolution.
iii. OSSs were not capable to support service restoration vs. technology repair.

Performance Management:

i. KPI reporting was not based on peak traffic (95 percentile).
ii. Inadequate ETE dashboard for customer/network-facing performance management.
iii. Monitoring activities focused on outages, and not on service degradation.
iv. Lack of ETE visibility into network services-EMS (focus on network elements) vs ETE OSSs (QoE) management of the network.

3.3.1.2 Assessment of Network Domain for QoE

Next, we looked at the performance contribution to the QoE by each component of the network footprint (Capacity and Performance Management: Best Practices White Paper, 2005) and (Balakrishnan, 2009).

To begin with, it is critical to have the QoE performance data cascaded down as targets/benchmarks to each component of the ETE network.

As we stated before, most of the CSPs did not have a real-time QoE performance monitoring dashboard in place, nor the targets for each component of the network footprint. We assumed a max 5000 kilometers for On-Net connectivity and > 5000 Kilometers for off-net. Here is what we used as the benchmark for the assessment of the performance of the components of the network Figures 3.7 through 3.10:

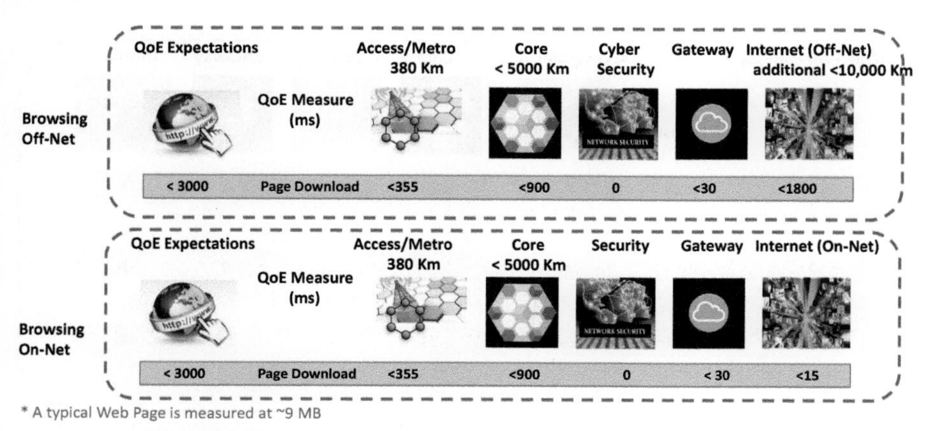

* A typical Web Page is measured at ~9 MB

Figure 3.7. QoE performance benchmark for browsing typical web page*.

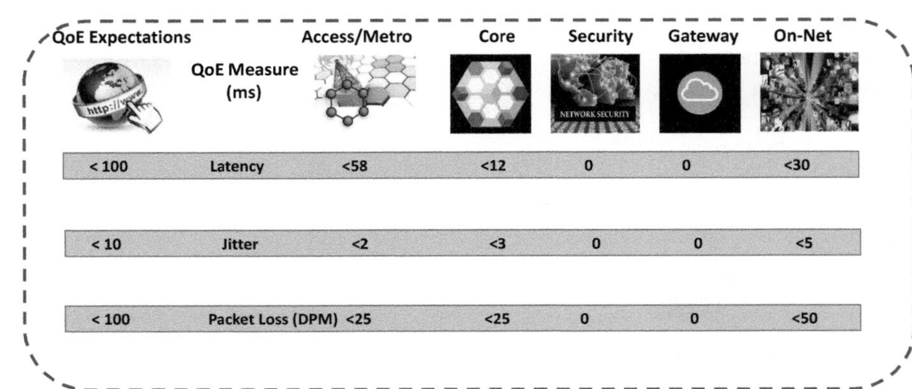

Figure 3.8. QoE performance benchmark for streaming (on-net).

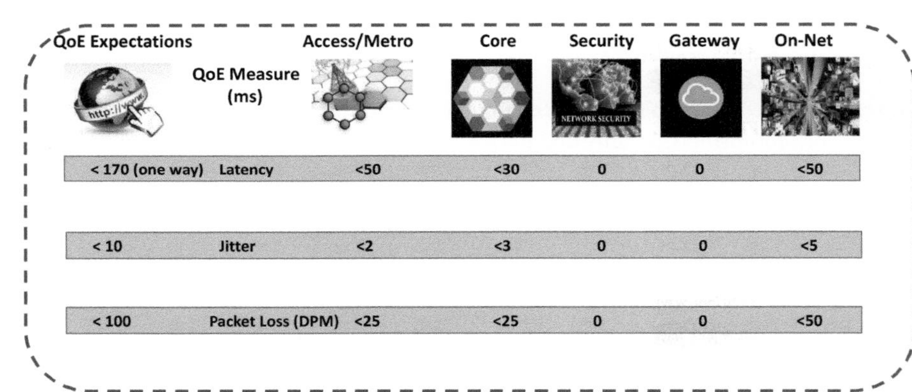

Figure 3.9. QoE performance benchmark for fo VoIP (on-net).

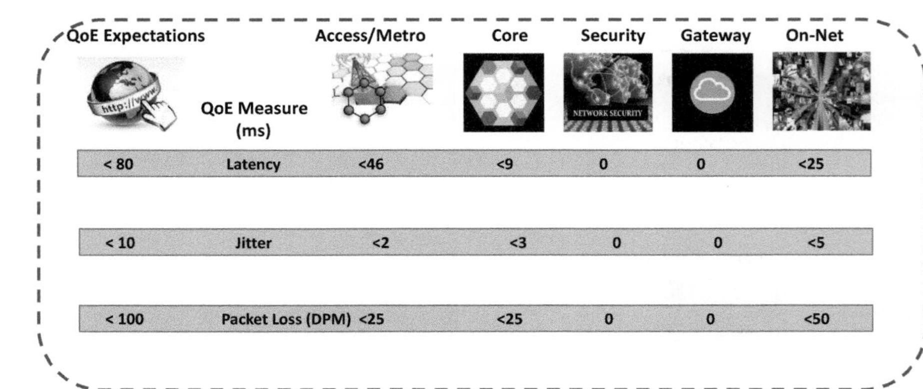

Figure 3.10. QoE performance benchmark for interactive video (on-net).

Here is what we learned from the actual performance of the many components of the CSPs' networks during peak traffic time and/or under failures, before the transformation Figures 3.11 through 3.14).

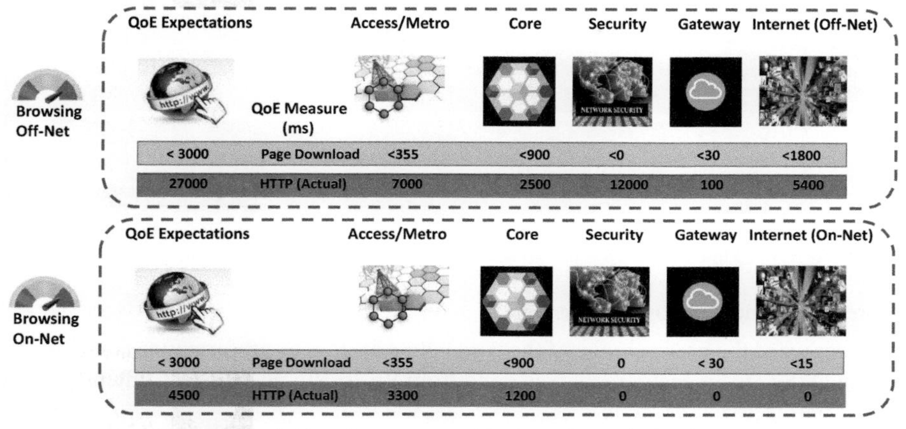

Figure 3.11. QoE performance actual for browsing typical web page.

VoIP

QoE Expectations	QoE Measure (ms)	Access/Metro	Core	Security	Gateway	On-Net
< 170 (one way)	Latency	<50	<30	0	0	<50
305 (one way)	Latency	130	50	0	0	130
< 10	Jitter	<2	<3	0	0	<5
15	Jitter (ms)	<5	<5	0	0	<5
< 100	Packet Loss (DPM)	<25	<25	0	0	<50
25000	Packet Loss (ms)	16500	890	0	0	7610

Figure 3.12. QoE performance actual for VoIP (on-net).

Streaming

QoE Expectations	QoE Measure (ms)	Access/Metro	Core	Security	Gateway	On-Net
< 100	Latency	<58	<12	0	0	<30
194	Latency (ms)	<157	<37	0	0	0
< 10	Jitter	<2	<3	0	0	<5
10	Jitter (ms)	4	3	0	0	3
< 100	Packet Loss (DPM)	<25	<25	0	0	<50
81000	Packet Loss (ms)	75000	6000	0	0	0

Figure 3.13. QoE performance actual for streaming (on-net).

Interactive Video

QoE Expectations	QoE Measure (ms)	Access/Metro	Core	Security	Gateway	On-Net
< 80	Latency	<46	<9	0	0	<25
< 194	Latency (ms)	<157	<37	0	0	0
< 10	Jitter	<2	<3	0	0	<5
10	Jitter (ms)	4	3	0	0	3
< 100	Packet Loss (DPM)	<25	<25	0	0	<50
81000	Packet Loss (ms)	75000	6000	0	0	0

Figure 3.14. QoE performance actual for interactive video (on-net).

Core Backbone Network Assessment:

We measured the performance of the core networks for the performance contribution to the QoE. This measurement was done from PE-to-PE, and PE-to-Content (on-net & off-net) (Kirstädter, 2006).

The CSPs that we transformed often had multiple independent core networks. In many cases, independent networks were the cause of acquisition activities, in other cases the legacy technologies were a driver for the creation of multiple data networks, each serving a specific customer space. Each network was customized to serve a unique customer base, often business customer centric.

For example (Figure 3.15), we could identify several data networks in place with ATM and FR technologies to serve business customers as well as backhaul for the consumer and business xDSL and wireless, several IP networks to serve consumers with internet access, business Managed Internet Services (MIS), Hosting and IPTV, legacy circuit switch networks for voice and low-speed data services, and of course more IP networks for mobility customers, etc.

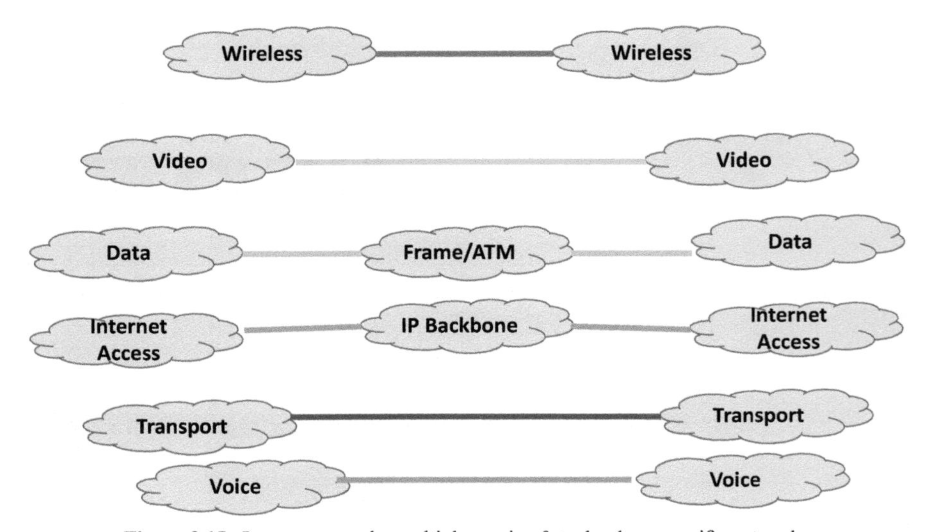

Figure 3.15. Legacy networks-multiple service & technology specific networks.

This arrangement by the CSPs was constructed over many years. Time after time, the CSP had to address the acquisition activities and respond to the need of the customers. Furthermore, given the state of the networking technologies, a new network had to be built which forced redundant services to be built for each network, redundant network care activities, capacity management, provisioning, call centers, maintenance, billing, customer care, etc.

Also, more and more of the CSPs' acquisitions in recent years have contributed to numerous access, aggregation, core, content networks that are now owned by the parent CSPs. This situation is begging for efficiency, both CAPEX and OPEX, through a unified IP network and having all customers migrated to the ETE One-Network and One-Service-Platform.

When we did the math on the business case for a convergence of wireline/wireless/video networks, the conclusion was an astounding CAPEX and OPEX reduction over 70%, with most in-year ROIC (Return On Invested Capital), and assurance for a BII QoE.

In all assessment cases,we were surprised with the poor performance for the core networks. We documented a 2.5 seconds actual performance to the browsing experience compared to the benchmark of 0.9 seconds during the peak hour. The jitter was significantly above the target of 1 ms, and the packet loss was over the target for 25 DPM. In many cases,the Core network demonstrated up to 300 ms variation in latency under failure of one fiber span, or one Node. It also showed significant latency under sunny-day operation, up to 200% of normal latency given the geographical size of the core network.

Overall, this indicated inadequate KPI reporting, which was based on the Day-Average measurements and not the peak traffic at ~ 95 percentile. We also learned that the core network was not designed and built for a delay (round trip time) of 12 ms per 1000 kilometers, during peak hour, and it demonstrated that the core was unable to support the high availability target of five 9's.

In all cases, the flow of the traffic, to and from the content, was not optimized for the core, which impacted the performance of the service flows, as well as the inefficiency of the core capacity.

(a) On-net traffic: In all cases, on-net content was being dragged from the on-net sources (CDN) behind the remote Access Routers through the core to the regional PEs serving area, and through the aggregation and access network to the eyeballs. This arrangement was the main contributor to higher latency, and significantly higher capacity consumption in the core.

(b) Off-net traffic: In all cases, off-net content was being dragged from the on-net sources (CDN and/or data centers), behind the Access Routers (not the Peering Routers) through the core to the GW routers, and through the other CSPs, to the eyeballs on their networks. This arrangement was the main contributor to higher latency, and significantly higher capacity consumption in the core (expect > 50% network port cost reduction for the off-net traffic by placing the off-net content behind the peering routers).

Due to simultaneous service degradation, under failure, for browsing, streaming, IV, and VoIP, it was clear that there were no CoSs implemented in the IP cores, nor was there bandwidth guaranteed for real-time sensitive and premium services. The core was often congested; protection capacity was hardly in place when needed during network failures. Capacity in the core was managed assuming bursty traffic pattern vs streaming traffic which requires a constant bit rate. The root cause was associated with the assumed inaccurate contention ratio (contention ratio must be ~ 1:1 for constant bitrate contents) that was used to manage the capacity in the access/aggregation/core. Also,an adequate real-time traffic steering mechanism not in place to prevent local congestion, and was not architected with the right number of nodes and meshing. The core was built for the sunny day operation, and it was contributing to the poor QoE by stealing a much-needed latency budget from the access part of the network. It also demonstrated that the capacity expansion may not have been driven by the sound engineering rules.

We also did many assessments of the services flowing through the network infrastructures for L3 vs L2. The purpose was to ensure if there was an opportunity to redirect a service flow through a more unit cost-efficient network infrastructure. For example, in the layer 3 services, such as IP/MPLS VPN, VPLS, the Internet must flow through the layer 3 network infrastructure, at the same time, services such as metro/inter-city ethernet, connectivity to data centers, Cell Site Backhaul, WiFi backhaul, point-to-point transmission and transit traffic should be flowing through a lower cost CAPEX infrastructure. In many cases we identified up to 5X traffic that could have been offloaded from the L3 through a lower unit cost CAPEX transport infrastructure.

3.3.1.3 Access/Aggregation Network Assessment

Here, we measured the performance of the access/aggregation networks for their performance contribution to the QoE. We measured the performance from the customer premise-to-access node, -to-aggregation nodes, and -to-the PE node for both business and consumers. We also measured the performance of the Private Line services (P2P) for business customers. This measurement was done for downlink traffic as well as uplink.

The CSPs often had multiple independent networks. The legacy technologies were a driver for the creation of multiple access/aggregation networks, each, serving a specific customer space. Each network was customized to serve a unique customer base for business (e.g., TDM/EoTDM, Sonnet backhaul) and consumers (e.g., 3G/4G, xDSL, ATM backhaul).

In all cases, we documented 3–7 seconds actual performance for the browsing experience compared to the benchmark of 0.3 seconds during the peak hour. We measured latency (xDSL/3G/4G) of ~ 140 ms/160 ms compared to the benchmark of 50/40. For xDSL, jitter was a bit above the target of 2 ms, and the packet loss, in many cases, was up to 30,000 DPM vs the target for 25 DPM.

In many cases, the access/aggregation network demonstrated service disruption under failure of one fiber or node. It also showed significantly more latency on the uplink for consumers under sunny-day operation.

Overall, the performance assessment for the access/aggregation indicated:

(a) for business customers, there was a plethora of Single Point of Failures in the access/aggregation network,and it demonstrated that the access/aggregation was unable to support high availability target of five 9's. The P2P transport services were not provided on the shortest path, nor were the P2P protected connectivity services on the same distance path. In many cases, we could document a back-up path half an order of magnitude longer than the original path (which proved that the network was not designed and built for delay of 12 ms per 1000 kilometers, during peak hour); the network was often congested, which was one of the root causes for significant packet loss impacting UDP services such as streaming/VoIP.

(b) the same issues for the business customer were also impacting the consumers, plus the fact that the limited uplink of the last mile for the xDSL/GPON was a bottleneck for the ability to offer digital services requiring symmetric bandwidth, such as security and surveillance.

(c) For business and consumers, CSPs, across the world, have built their networks based on the "Asymmetric Communication" technologies for consumers, and a "Symmetric Communication" for the business. The fundamental assumption for consumers is that the downlink speed is significantly higher than the uplink speed. As a result, the spectrum allocation (e.g., in DSL, GPON, wireless) is designed and deployed with this mindset for consumers. This approach has forced CSPs to deploy two different sets of high unit cost-intensive platforms for consumers, and separately for businesses (requiring separate packet processing, fiber connectivity, call centers, etc.).

3.3.1.4 Content Network & IXP Assessment

Here, we measured the performance of the content networks for their performance contribution to the QoE. We measured the performance for on-net traffic, from PE-to-CDN/IDC, and for off-net traffic, from PE-to-IXP-to-CDN/IDC.

The CSPs that we assessed, often had exposures to CDN/IDCs, however, the network performance was poor. In that, the placement and the architecture of these platforms were questionable and far from being optimized.

In all cases, we documented up to 6 seconds actual performance for the browsing experience (for the off-net content) compared to the benchmark of 1 second during the peak hour. Often there was no load-balancing architecture in place, nor an iCDN (Intelligent CDN) to recover from the congestion in-network or content-node. The jitter was a bit above the target of 1 ms, and the packet loss, in many cases, was in the expected range of < 10 DPM for off-net/on-net content.

In many cases the content network demonstrated service disruption under failure of one fiber connectivity to the content, or the entire node.

(a) On-net traffic: In all cases, on-net content was being dragged from the on-net sources (CDN) behind the remote Access Routers through the core to the egress PE, through the aggregation network and the access to the eyeballs. This arrangement was the main contributor to higher latency, and significantly higher capacity consumption in the core.

(b) Off-net traffic: in all cases, off-net content was being dragged from the on-net sources (CDN) behind the Access Routers (not the Peering Routers) through the core to the GW routers, to the cooperating networks, and to the eyeballs on the other networks. This arrangement was the main contributor to higher latency, and significantly higher capacity consumption in the core (expect > 50% network port cost reduction for the off-net traffic by placing the off-net content behind the peering routers).

3.3.1.5 Network and Users Security Assessment

The subject of the network and security will be addressed in the next release of this book. The intent of the network security must be focused on making the network assets invisible to hackers. We evaluated many of the CSP networks in order to establish how they use real-time data and deep-packet inspection to direct the traffic to a look-alike network, to look for patterns of attacks—to get into the minds of hackers and stop them before they launch. This knowledge mining is mastered with few CSPs that grasp the concept and apply the principles effectively. Government assets may need to use this technique to prepare for future cyber-attacks, an effort that will require federal legislation. This is the only way to avoid major cyber wars in the 21st century. If not addressed, the negative impact of cyber wars could be colossal for humankind.

3.3.2 Assessment of Network-Facing Dashboards

In this step of the assessment, we strongly recommend to evaluate the specific dashboards, as well as the drill-down capability from the QoE to specific network element that is impacting the QoE performance. Here we assessed 4 key dashboards: competitive, executive, operations management, and working level. Also, we examined the drill-down capability from the QoE to the Working Level dashboard.

3.3.2.1 Assessment of CSP Competitive Dashboard

This competitive dashboard is essential to ensure that CSP is executing to maintain a competitive position in the market place. In this examination, we are looking to see if the competitor's performance is being monitored? If the CSP's performance is being captured and presented to enable an apple-to-apple comparison, and finally if the business targets, for the CSP, are established and used to drive to a competitive position.

Today's tier 1 CSPs, unlike tier 2 and below, are using an example of the following competitive dashboard (Figure 3.16) to assess their overall competitive performance (RootMetrics, 2020).

Often, we learned that such a dashboard did exist, however, in many cases, where it was in place, it was not reflective of the peak hour performance, nor it was reflective of the performance of the Macro Service Flows. As a result, it produced miss guided confidence about CSPs' performance, who were measuring their QoE performance through the day-average approach vs peak-traffic-hour.

Market	Service	Service Type	Own Performance	BII Performance
Wireless	**Data**	Speed (DL/UL)	5.3 / 4 Mbps	20/9.5
	Voice	Call Drops	1.3%	0.2%
	Text	Texting within 10 sec	98.3%	99.1%
Wireline	**Internet**	Speed (DL/UL)	6/1 Mbps	100/10
	Biz: Large Enterprise	CoS (latency RTT)	75 ms	20
		CoS (Jitter)	45 ms	15
	Biz: Small / Meduim/ Large	Service Availability	96.1%	99.999%
	Strategic Services	Content, VPN, L2 E, Wavelenght, PL, Unified Comm, Hosting, Security, Cloud	3 of 8	7 of 8
		National Churn Rate:	**1.93%**	**<0.8%**

Figure 3.16. CSP competitive dashboard.

Another finding was that the CSPs' competitive shortcomings were not consistently used as a driver to establish the goals for the business. The Competitive Dashboard must be used as a driver to help in understanding the competitive landscape, and it must be internalized as a part of the goal-setting process.

Also, we look into the churn rate and examine whether the CSPs had a process in place to mine insight into the churn rate. A high churn rate relative to the competition is indicative of a myriad of problems. We look at the processes that were used to understand the reasons for the churn. The root cause for the churn rate is key to understand, and often is the outcome of customer interviews when they chose to switch to a competitor, and/or driven by the shortcoming in measurement and internalization of the QoE. In most cases, CSPs were attempting to get to the root cause, but when we evaluated them, their findings proved useless and misleading. For example, for wireline consumers, the real issue was the poor browsing experience during peak hours, not the day-average. In other cases, the poor gaming experience was the real issue, while the reported competitive KPIs were based on the day-average indicated an "ok" status. For wireless customers, it was the poor QoE experience, but, most importantly, it was a function of the location of experience, e.g., "at home", "in-building", "Driving local", and "traveling", that was contributing to the churn rate. We looked at the reasons for the churns. This included a series of issues that must be quantified and understood, including the signal quality, dropped calls, poor sound, VMS notification delay, problems sending and receiving text, difficulty connecting to internet, slow browsing experience, and the better pricing alternatives offered by the competition, etc.

We learned that most influencing reasons for the churn rate were related to the "at home experience" and the "driving local". However, this fact was missed from the customers' interviews.

For many of the CSPs that we assessed, the root causes for the churn were numerous, including poor coverage, inadequacy of the network capacity, especially during peak hours, far-reaching content, and for the delay in call-set-up time, as it was often designed for a low bandwidth control channel that causes congestion during call set up time.

3.3.2.2 Assessment of CSP Executive Dashboard

The executive dashboard (IBM Business Analytics for telcos, 2010) is aimed at ensuring the calm that must exist over the operation of the ETE network. This dashboard ensures that:

(a) The preventive actions (such as a simple change of air filter on a router, or a software scan for the latest software version) are being taken at the right time to avoid network and service incidents.

(b) The proactive actions (such as potential capacity exhaust in a POP) are rule-based and being executed on time, in anticipation of the customer demand, and,

(c) The predictive actions (such as the deployment of an optical amplifier to alleviate fiber degradation) are being pursued on time to avoid long cycle corrective actions.

The executives must be engaged across the board, to monitor, and in case of potential systemic performance issues against the targets, be prepared to marshal company-wide resources to fix the problems proactively, before they turn into a systemic issue and a disaster.

Disaster is when you have to gate the global sales operations due to lack of scalability and the supply-side issues, or to dance around an unpredictable sales uptick, which is often the case. Disaster is to face a global SLA compliance issue such as latency as it is impacting global customers. Disaster is to face a systemic QoE issue in one or more operating regions.

According to our assessment, today's CSPs' executives often delegate this important task to their operations managers. As a result, CSPs have a distributed view of the performance of the overall network, without a consistent set of rules applied to flag the performance issues and the proactive actions that must be taken promptly to avoid the disaster.

Our assessments often indicated that a comprehensive Executive Dashboard, as stated below (Figure 3.17), was missing. This dashboard is to alert the executives for any deviation from the performance targets, through R/G/Y, and enables the

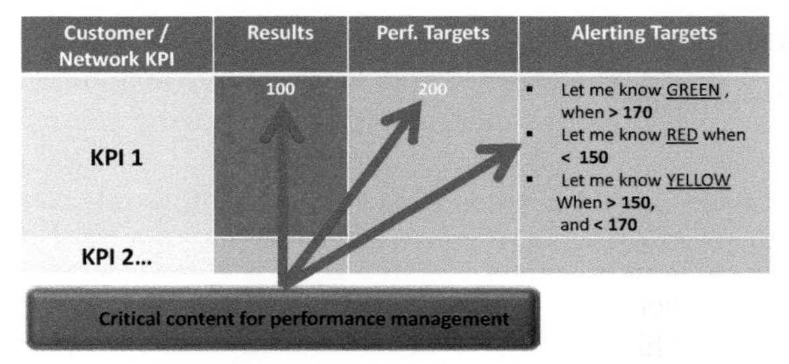

Figure 3.17. DSP executive dashboard.

executives to ensure that network and services are being cared for in a proactive, preventive, and predictive way.

3.3.2.3 Assessment of CSP Operations Management Dashboard

The Operations Management dashboard is aimed at ensuring that each domain of the ETE network is being cared for in a consistent and well-defined way to be preventive, proactive, and predictive.

The Operations Managers must be engaged across the board, to monitor, and in cases of potential performance issues against the targets, get engaged to help the process associates to fix the problems proactively, before they turn into a disaster as described in the previous section.

Operations Managers must be focused on ensuring that the team responsible for the performance of the domains of the network are taking the appropriate actions, proactively, to ensure that the overall network performance is not compromised. Here, we are looking for actions that are being taken within the allowable time windows to proactively care for any potential issues with the performance of the QoE, the network, and the services.

As we assessed many CSPs' Operations Management dashboards, we concluded that the dashboards they were using, were inadequately equipped with the "Alerting Targets", and the "Action Targets" (as depicted in Figure 3.18). As a result, it was left to the seniority and the experience of the process associate to decide when it is the right time to trigger an action to fix the network and services issues. This situation was presenting a significant challenge for a consistent treatment of the issues at a global stage.

Working Level KPIs

Customer / Network KPI	Results	Perf. Targets	Alerting Targets	Action Targets
KPI 1	100	200	▪ Let me know GREEN, when **> 170** ▪ Let me know RED when **< 150** ▪ Let me know YELLOW When **> 150**, and **< 170**	▪ Augment capacity at 160
KPI 2...				

Figure 3.18. DSP operations management dashboard.

3.3.2.4 Assessment of CSP Working-level Dashboard

The working-level dashboard is key in enabling the process associates to make real-time decisions and care for the health of the network. This dashboard must be linked up to the QoE dashboard.

Here, we examined the availability of the dashboard with the focus on presenting the process associates with the facts on the targets, actuals/results, alerting targets, and any appropriate actions that are taken and documented to proactively address the potential issues. This dashboard must be linked up to the QoE dashboard. The drill-down capability from the QoE dashboard to the Working Level Dashboard is essential in maintaining calm and control (Figure 3.19).

Looking at many CSPs, we established that all had implemented some sort of dashboard. However, in all cases we could not see the linkage from the QoE dashboard to the working level dashboard and the actions that had to be taken by the operations to fix the QoE issues. Likewise, we could not identify the presence of rule-based decision-making to fix the performance and operational issues, i.e., minimal alerting targets, no action targets, and nowhere to systematically capture the actions that were taken by the process associates to fix the operations issues.

Next, we looked at the ETE network operations to examine the extent of a customer-centric approach to operations. We found that the handling of the service outages was time-consuming, sometimes taking days. For example, in the case of a typical DHCP/inbound mail exchange outage, the root cause revealed that the focus of recovery was based on repairing the failures and not the service restoration. We also noticed that, in many cases, the services were being disrupted with a simple outage, i.e., the network was unable to heal itself.

In many cases, when we root-caused the service outage, it was established that the network configuration was changed by the R&D personnel without any oversight provided by the operations and/or NOC.

In many cases, we sized the customer disruption for a typical "Node Move", where the customers were being moved to another access component in order to alleviate the access congestion, at 100,000 DPM, which proved the process was out of control.

In a nutshell, the facts presented MTTR (mean time to repair) of 7–10 hours, TTR (time to restore) of 50% within 2 hours vs 90%...outages that were alerting Senior Leadership Team (SLT) notification vs SLT engagement, lack of Service Methods of Procedures (SMOP), high number of POE every week of 40 cases vs zero, unlimited Root Access by non-production personnel, network not driven by engineering rules, and on and on.

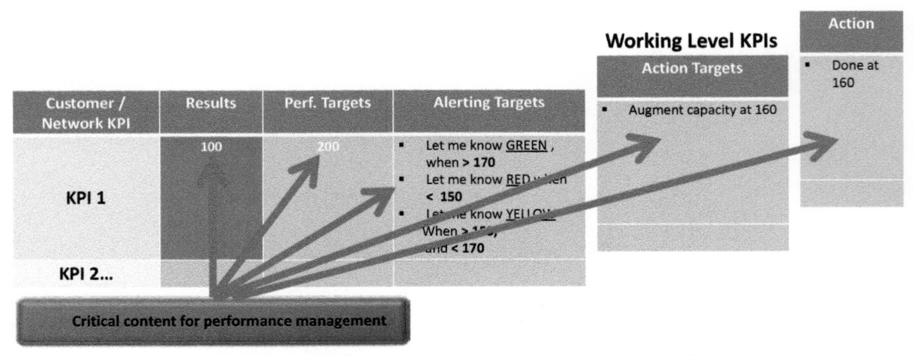

Figure 3.19. CSP operations working level dashboard.

3.3.2.5 Assessment of Dashboard Drill-down Capability

Finally, we looked for deviation in the QoE performance, and the correlation with the drill-down capability in order to narrow the root cause for the service degradation down to a specific network element.

In most cases, the CSPs did not have a QoE dashboard. In the few cases where QoE was in place, there was a disconnect between the QoE dashboard and the operation dashboard,in that the operations dashboard was often structured to manage network outages rather than providing insight into the QoE performance and service restoration.

We have led transformation efforts where, in a few cases, the QoE dashboard was operational, and the drill-down capability for the root cause was in place; and in many cases where this dashboard was not available. Also, we have been to other cases where the dashboard was in place, but not measuring the QoE performance properly through 95 percentile peak traffic.

Here is one example, from a few CSPs with the QoE dashboard, where there were no drill-down capabilities (Figure 3.20) to establish the root causes for the poor QoE performance during the peak hour.

Based on our assessment, the key issues were: (1) the QoE performance was poor, (2) the QoE targets were not driving the network design for the key services, and (3) there was no drill-down capability to establish the root causes, i.e., the working level dashboards were useless and isolated from the customer experience.

The QoE drill-down capability must point to the root causes for poor performance: poor resiliency, congestion, security on the critical path, long reach internet content, poor access/metro performance and bandwidth bottleneck. Also, this drill-down capability must establish if there is a central planning organization, responsible for enforcing performance targets for the ETE components of the network.

In many cases where the drill-down capabilities were not available, we had to implement temporary sensory devices throughout the network to enable data extraction and data mining, establish performance targets for each component of the ETE network, and enable determination of the root causes, leading to the optimization and/or redesign of the DSP platform to deliver 12 ms latency (RTT) per 1000 kilometers, to redesign the location of the content relative to the eye-balls, to establish a floor level bandwidth on the access network to ensure QoE for browsing customers, and/or HDTV/4K content customers, etc., we also had to create multiple scenarios to measure the actual performance under failure as well.

If this dashboard does not exist, you must be prepared to build such a dashboard in a matter of weeks. This dashboard with the drill-down capability is the key tool to use to develop a fact-based understanding of the effectiveness of the architecture, implementation, and management of the network.

3.3.3 PMO: Assessment of Network & OSSs/BSSs Architecture

[This section will be covered in the next release of the book. It will cover the specifics for the following subjects: Service over IP OSSs/BSSs, Customer/Vendor/Partner Facing OSSs/BSSs, and Network Facing OSSs].

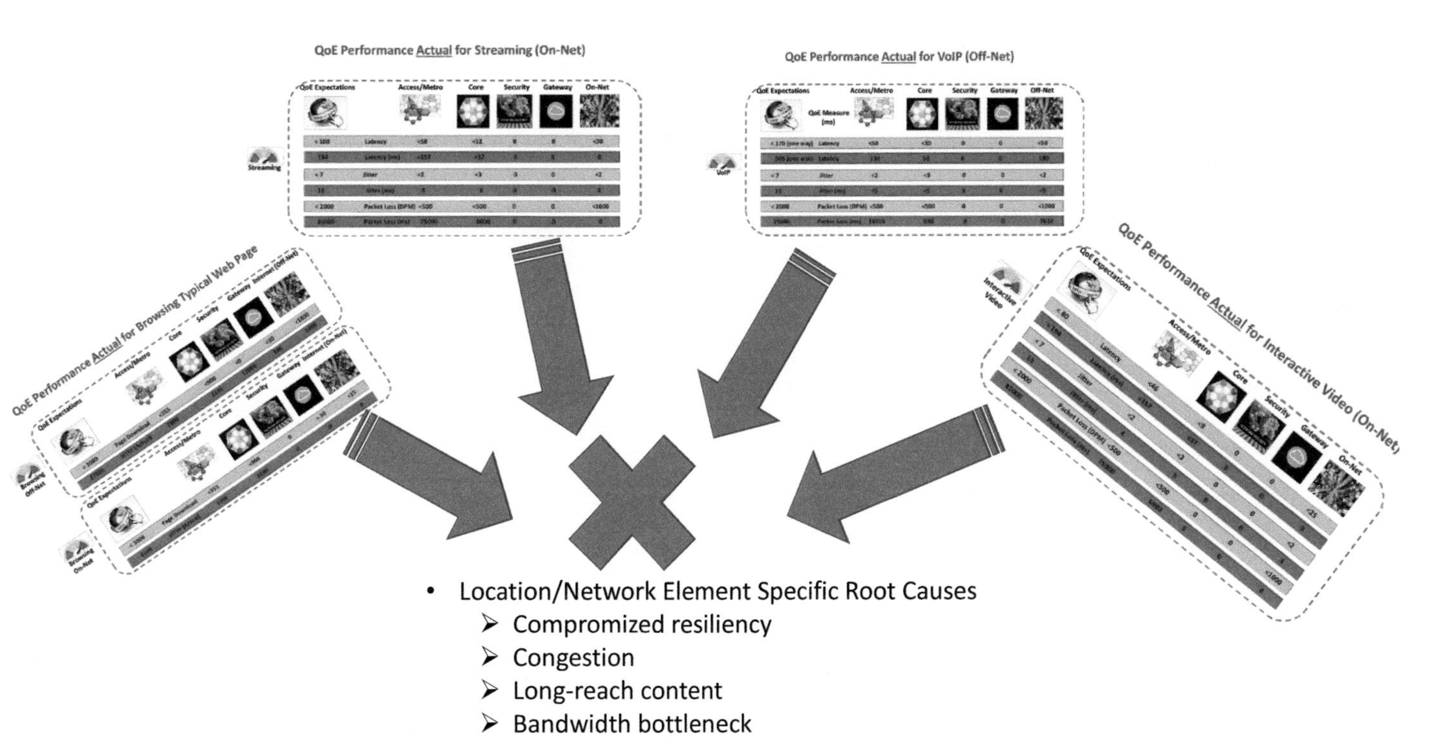

Figure 3.20. Lack of laser sharp drill-down from QoE to the root causes.

3.4 PMO: Assessment of CAPEX Unit Cost

Management of the CAPEX (Meakin, 2012) (Monetizing investments in telecom network infrastructure, 2018) is the most challenging aspect of the CSPs' CFOs job. The CAPEX ratio is to be maintained at the right level for a sustainable CSP. The sweet spot for this ratio is between 12% and 18% depending on the 3-play/quad-play connectivity service provider status. However, this ratio will be changing to some extent as the traditional CSPs are expanding into high margin digital services as they achieve the DSP status.

Looking at the global revenue growth vs traffic growth, it shows that while annual revenue has grown at ~ 3–6% from 2010 to the present, the data traffic is growing at greater than 40% YOY. With a closer look, it can be seen that the access bandwidth (wireless and wireline) is growing at ~ 30% YOY, while the IP backbone is growing at ~ 50% (the higher the number on the backbone is due to the "via or transit" traffic). This growth is primarily driven by video and high speed enterprise data applications.

At the end of the day, The CSPs need to add network capacity to manage the growth in the traffic. As we all know, any increase in network capacity drives CAPEX. CAPEX expenditure impacts the EBITDA Margin and the Free Cash flow (i.e., EBITDA minus CAPEX). This CAPEX expenditure is spent in two forms: (1) CAPEX for running the business, and (2) CAPEX for improving the business. The first is the CAPEX required to support the traffic growth leveraging the existing, approved-for-use technologies, and the second is the CAPEX required to build new technologies to replace legacy technologies.

In other words, any increase in CAPEX is best to be mitigated by the increase in revenue. The less the network traffic growth is captured on the high CAPEX unit cost of the "legacy network", the more favorable free cash flow is for the CSPs.

The BII practice to calculate the CAPEX Unit cost is to divide the total capital spent in each year by the total incremental volume (e.g., Gigabit Ethernet consume) in the same year.

The CAPEX reduction focus must be on the growth CAPEX (i.e., the "running the business" to capture the traffic growth, leveraging the approved-for-use (AFU) on new network technologies). The replacement CAPEX (relatively in small amount compared to the CAPEX for "running the business") leverages transformational technologies and is key in enabling the optimization of the CAPEX to support the traffic growth going forward.

Let us repeat one more time, it is of utmost importance that the growth CAPEX must be spent on the lowest unit cost technologies. However, it must be understood that the growth CAPEX unit cost reduction is enabled through the replacement CAPEX. The key is the timing for the CAPEX spend on the replacement capital, and the growth capital. If the timing is near perfect, then the traffic growth will be captured on the most CAPEX unit cost-effective platforms.

All CSPs are mindful of the mismatch of the incremental YOY revenue growth, and the high unit cost of CAPEX which is negatively impacting the EBITDA Margin and the Free Cash flow. This issue has to be addressed through network transformation.

In this analysis, we must consider CAPEX to exclude licenses and spectrum fees. The CAPEX unit cost analysis must focus on the capital view, including the OSSs/BSSs. The unit cost analysis within this view must be used to determine where capital investment can best improve overall capacity efficiency.

The bottom line for a high-performing, CAPEX-efficient network architecture is the cost of the lit ports in the content data centers, access, aggregation and the core. This "Lit port" concept is inclusive of the routing/switching/RAN technology as well as the transmission connecting and activating the ports.

A typical high performing network could be deploying 10's of backbone routing nodes, with many routing chassis in each routing node, up to 100's of routing chassis, 1000's of aggregation nodes, 10,000's of access nodes (for wireless and wireline), and 1000's of CDN nodes. The majority of the cost is in the Access nodes and the cost of access "lit ports".

Given all this, you need to look at the unit cost for the access and aggregation for the data services (e.g., VPNs, HSIA), then develop a model for key ETE connectivity services. This model must capture the unit cost structure for these segments of the network: customer premise-to-serving POP, from serving POP-to-regional aggregation POP, from regional aggregation POP to national aggregation POP, from national aggregation POP to Core ingress backbone POP, and from Core ingress backbone POP to core egress POP, and finally to the Content.

The service unit cost must be constructed for every legacy network. Any major CSP has many types of networks in their network portfolios, including TDM/Data wireless networks (for voice and data), TDM wireline networks (for voice and low-speed data), Legacy wireline Data Networks (for High-speed internet access and data services), Video Network (for IPTV), and IP/Ethernet Networks (for all applications).

The key and fundamental question: what is the ETE CAPEX unit cost for the generic connectivity services, as well as the unit cost for the content platforms? We need to build the ETE unit cost model for 3 major generic connectivity services that are provided through the above networks: (1) Mobility service (measured by equivalent million minutes of usage), (2) Data service (for consumer/business customers, measured by equivalent GigE), and Video (for consumers, measured by Living Unit Passed).

Here is an example of the model for these generic services (Figure 3.20.1), from the customer premise (consumer/business/cell site), through the core to the variety of destinations (e.g., internet, IDCs, on-net termination, off-net termination). This model is essential to understanding where CAPEX is being spent and what areas of CAPEX expenditure present the most value from a transformational perspective.

Next, we should establish the peak load (Gbps) for each one of the CSP's networks, and add them up to get the size of the peak load for the entire PMO network for the CSP.

This total peakload is what is growing at a rate of 50% YOY, and the on-net access for the CSP is growing at 30% YOY.

Next, we should capture the demand growth on the legacy networks. In many cases, we identified the following 4 types of legacy networks and assessed the impact to the Free Cash Flow (EBITDA minus CAPEX):

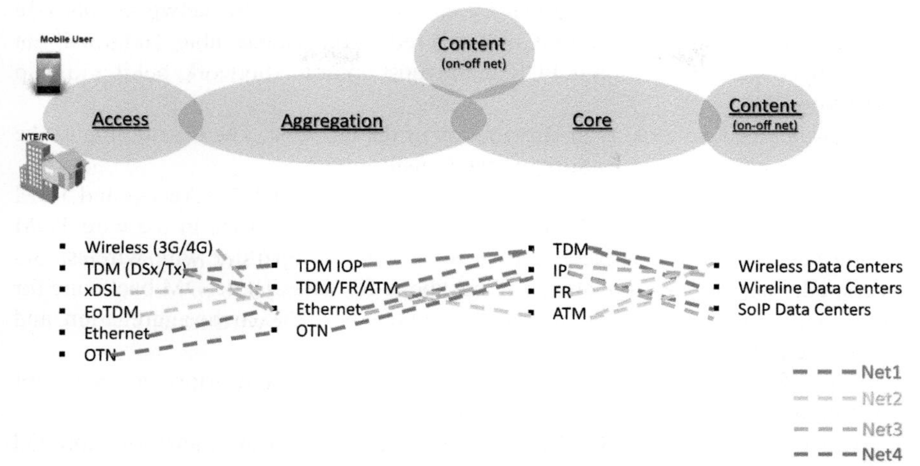

Figure 3.20.1. ETE capital unit cost.

Net 1 (TDM Low-speed connectivity: TDM access, T1 DS0 Access, 3G wireless access, TDM IOF &Agg, Sonnet Transport, TDM core, Content platform). The unit cost for an equivalent 1 GE ETE connectivity to the content was 40X over Net4—the most CAPEX unit cost-efficient network

Net 2 (Data low speed connectivity: EoTDM access, 3G+ access, FR/ATM aggregation, Sonnet Transport, FR/ATM core, Content platform). The unit cost was 30X over.

Net 3 (Data/Legacy+ connectivity: 3G/DSL/EoTDM access, low-speed ethernet aggregation, DWDM Transport, IP core, Content Platform). The unit cost was 6 times over.

Net 4 (Data- High-speed all IP: 4G/xPON access, scalable Ethernet aggregation, DWDM (80+ channel)/OTN transport, scalable IP Core, Content Platform). The unit cost was 1X for an equivalent 1 GE connectivity service.

At this stage, given the insight into the above four nets, we can then use the typical global tier 1 CSPs for the assessment of their Free Cash Flow, if they were to capture the traffic growth on each Net as described above.

(a) When we developed the model to capture the traffic growth on Net1, the incremental CAPEX requirement forced an FCF (Free Cash Flow) drop by as much as 70%; this was a kiss of death as the dividends would suffer immensely.

(b) In doing so with Net2, FCF dropped by as much as 50%. An unacceptable dividend impacting scenario.

(c) For Net3, the FCF dropped by 15%.

(d) For Net4, the FCF dropped by 5%.

The bottom line is that it is critical to transform the ETE network, and to time the Replacement CAPEX to get the transformed network ready to support the traffic growth on the lowest CAPEX unit cost network.

Also, it is important to note that by consolidating all CSP networks into one unified network, additional CAPEX unit cost reduction is achievable, and it will not only compensate for any drop in FCF on the most optimum network, but it will help to improve FCF.

Now, let's take a look at the distribution of the unit cost. The distribution of the CAPEX unit cost is key in prioritizing the transformation.

For Net1, 62% of the ETE CAPEX unit cost was in Lit TDM Access and TDM Access backhaul, 25% in TDM aggregation, and less than 13% in the core TDM routing and transport. For Net3, a case for a wireless operator with wireless 3G access, ethernet aggregation, and an IP backbone for data and a TDM backbone for voice. In this case, 58% of the unit cost was in the RAN, 26% in the aggregation, and 16% in the core.

Any CSP with one network of type Net1 or Net3 should prioritize access for transformation over aggregation and/or core.

However, many of the CSPs that we were engaged in the transformation had multiple networks with their independent core, aggregation, and access networks.

Overall, all individual networks demonstrated some ~ 60% of the unit cost in the access, close to ~ 25% in the aggregation, and the remaining ~ 15% in the core. For CSPs with multiple legacy networks, we established that the transformation and unification of the core is the most CAPEX unit cost-effective approach through full retirement of the legacy cores within-year ROIC. Also, we established that the cap-and-grow approach to the aggregation is the most effective investment approach to start the process of transformation with a multi-year objective to retire the legacy aggregation.

Access is the most CAPEX intensive transformation. Also, it takes many years to replace the legacy access with High-speed alternatives (through FTTP and/or/ wireless 5G+ technologies). The BII strategy for the transformation of the last-mile access is through a two-step approach: (1) Introduction of WiFi-offload through the WiFi Access Points in the high-density neighborhood, and (2) densification of the fiber in the neighborhood to pave the road for 5G+ wireless technology to enable High-speed mobility access.

Here is what to expect in terms of CAPEX unit cost reduction by moving to an all IP and ethernet, and 4G+ technologies: for the access lit port or equivalent One GigE, expect up to 90% unit cost reduction with scalability to 40+ Gbps over time leveraging EPON, WiFi-offload, wireless 5G+ technologies; for the aggregation lit port, expect up to 85% unit cost reduction with the scalability of an order-of-magnitude, leveraging Ethernet/IP technologies; and for the core lit port in the order of 80% unit cost reduction with scalability to 100's of Tbps per node through the unification of the core. These reductions are based on where the PMO technology is at, and the number of the legacy networks before the transformation.

3.5 PMO: Assessment of OPEX Unit Cost

Based on our experience, we can break the OPEX into the following components and their contribution to the overall cost: (a) Network Operations (~ 30%), (b) Customer care (~ 10%), (c) COGS (~ 20%), (d) Marketing, sales and admin (~ 40%).

High OPEX unit cost (Clinckx et al., 2014) is driven by the *defects* and the *cycle time* in the business processes (i.e., 80% of the OPEX) inclusive of customer-facing processes (such as marketing, contracting, ordering, provisioning, network equipment and facility inventory, billing, restoration maintenance, customer care) and network-facing processes (such as maintenance, development, etc.).

Through many assessments of the ETE CSPs' processes, we established that the cycle time is the root cause for the high OPEX unit cost. We have established that every 50% reduction in cycle time generates 40% reduction in OPEX. This cycle time reduction activity, when driven down to real-time performance, has produced > 80% reduction in OPEX unit cost. It enables > 98% flow through for service delivery, and help to increase the customer adoption of the e-servicing up to ~ 100%.

Defects:
To size the overall impact of the defects, we look at the CSPs' billing contra revenue relative to the total revenue. This provides an excellent source to quantify the defects in the overall CSPs' PMO customer-facing processes and systems.

The big picture of a CSP overall financial status requires that CSP establish a reserve, called Contra Revenue, to compensate for lost revenue due to errors in their bills for the customers, specifically the business customers. We defined "billing DPM" as a measure for the size of the defects that are generated in the contracting process, through ordering, provisioning, inventory management, and billing processes. It is the ratio of "Correct Charge Adjustment" to the "Total Billed Revenue", normalized to a million.

For example, based on our experience with the transformation, we have evaluated many CSPs with a substantial reserve set aside for Contra Revenue. Based on our assessment, we established the Billing DPM before the transformation from 70,000 to 100,000. In other words, some 7–10% of the gross revenue was being set aside in the "reserve" to address the errors in the bills.

The CSPs build their network for low DPM, it must be no surprise for the customers to expect the same performance in their bills. The billing DPM must be as close as possible to the engineered Network DPM. In this regard, a typical tier 1 CSP with a multi-billion-dollar revenue stream must be able to reduce its' contra revenue by 1+ billion dollars, and improve its' Free Cash Flow by > 20%.

Cycle time:
In many of the CSPs that we analyzed, a typical Connectivity Service Delivery cycle time (from ordering, to provisioning, to test-and-turn-up to billing activation) is measured in months and weeks for the business customers, or days for consumers. For the on-net business customers, the cycle time was in weeks, for off-net business customers it was measured in months given the complexity of the permit, ROW, and the build process.

We also looked at the E-bonding with the CSPs' partners for minimal equipment orders. It dictated a long cycle time, measured in days and weeks, to place an equipment order, receive the order and give acknowledgment to issue the payment.

Also, many of these CSPs did not provide self-servicing for connectivity services.

On the other hand, we found fault management on the critical path for service restoration. Often the services were disrupted during the time needed to repair the service, which took hours and often days.

3.6 PMO: Overall Gap Closure

Based on our collective professional experiences, we have completed the assessment of many CSPs with a laser-sharp focus on their poor performances. The following list is best aggregating the common theme for the sub-standard performances of many of the CSPs, as well as measured improvements to expect from the CSPs' transformation:

(A) **Low margin service portfolio and excessive churn:** limited service portfolio on connectivity services vs expansion to include services for digital services. Look for *up to 25% improvement in the EBITDA margins.*

(B) **Not competitive QoE Performance:** poor QoE is a *kiss of death for losing market share and increased and costly churn rate-look for 1% growth in earning for a reduction of 50% in churn.*

(C) **High CAPEX unit cost, operating multiple networks:** multiple redundant networks for wireless, IPs, TDMs, Data, voice. Anticipate *up to 80% CAPEX unit cost reduction by consolidating into one, scalable, unified IP Network.*

(D) **Not easy to do business with:** long cycle time in customer-facing processes, e.g., ordering, provisioning, restoration, as well as poor coverage. *Churn rate of up to 50% improvement, estimated equivalent to 1% of the increase in earning, and up to 80% reduction in OPEX when achieving customer self-servicing status.*

(E) **Inefficient OPEX unit cost structure:** drive defects to near zero with *up to 80% reduction in the OPEX unit cost.*

(F) **High cost of "reserve":** significant size of the "Contra Revenue" is what we need to go after. This is a low hanging fruit and it produces *up to 10% of the revenue in improving free cash flow and the margins*

(G) **High cost of SLA compliance:** expect*up to 5% of the revenue improvement in free cash flow and the margins*

Also, we have identified the root causes, each quantified for their impact, and identified specific Network Solutions to close the performance gaps. Below (Figures 3.21 through 3.26) is a summary of the gap closures:

CSP's Need	PMO Status	Network Solutions	FMO Status
High Margin Services ✓ Up to 25% improvement in EBITDA margin ✓ Enable gain in market share	●	• Original content & Subscription based services • OTT VOD, Music streaming, IPTV, Cloud based infrastructure, news/magazine subscription • iCDN,	●
Quad-Play Connectivity Services: ✓ Enable gain in market shares	●	• Foot print expansion: Video, Voice, Data, Wireless	●

Figure 3.21. Overall PMO gap closure.

CSP's Need	PMO Status	Network Solutions	FMO Status
Best-In-Industry QoE ✓ Must-do for a competitive and sustainable business	●	• Established high-performing performance targets based on QoE and SLA expectations • Five 9's service availability- Resilient Network/Services • ETE network built to support the 7 service flows during the peak hour & under failures • SLA Graded Service for Business • QoS / CoS / DS-TE • Manage to 95tile for Network KPIs in each domain • Strategic placement of iCDN / IXP • Core & Aggregation Designed for 12 ms Per 1000 Km • OSS Focus on Service Restoration and Service Degradation • Real Time SLA Management • Symmetric bandwidth Services • Operations discipline to run and operate the network: Customer communication, Ask YourSelf, Network Events, 3CP/MCB/TCB, Root Access, Outage management, M&Ps, Disaster Recovery	● ● ● ● ● ● ● ● ● ● ● ●

Figure 3.22. Overall PMO gap closure.

Need	PMO	Network Solutions	FMO
Services built on Low Capex Unit Cost ✓ Up to 40% reduction for One unified network ✓ Up to additional 70% reduction for scalable, All IP/MPLS/Ethernet/LTE+ network	●	• Unified-National Network for wireless/video/wireline ✓ Single L1-L3 network: DWDM & OTN/Ethernet/IP-MPLS ✓ Single L4-L7 service platform serving consumer/enterprise/wholesale/international: ✓ Single SoIP Platform ✓ Common Runtime Execution Environment ✓ One service logic for all access technologies: TDM/PSTN/UMTS/LTE/DSL/etc. • Shared network elements, facilities, and real estate: ✓ Integrated POPs: National, and Regional COs ✓ Integrated Data Centers: National/Regional, iCDN, Cloud • Scalable IP/MPLS Core, Aggregation, Multi-Service Access with ethernet uplink • One Access platform for Biz & Consumer • Preserve legacy access • OTN mesh transport for PTP and Access to L3 services • Efficient transport of low feature traffic, e.g., transit: thru circuit switch gateways, Session Border Controllers, and transcoding pools	● ● ● ● ● ● ●

Figure 3.23. Overall PMO gap closure.

Need	PMO	Business Process Solutions	FMO

Services built for low Opex unit cost
✓ Up to 80% reduction in Network ops, customer care, sales and admin.
	●	• Focus on Defects and Cycle time reduction throughout the processes
		• Establish Billing DPM as a measure for defect reduction: for contracting, ordering, provisioning, billing, and maintenance processes
		• Establish cycle-time= zero as a measure for improving customer-facing processes:
		  • Ordering, Service provisioning, and billing: Flow-thru with Point & Click real time
		  • Customer care: > 99% of calls resolved on the same call
		  • Service restoration: Zero time
		• Establish cycle time as a measure for improving network facing processes:
		  • Capacity planning & delivery: within days
		  • Mean-time-to-repair: 90% of cases within 2 hours
			●
			●
			●
			●

Figure 3.24. Overall PMO gap closure.

Need	PMO	Network Solutions	FMO
Manage Reserves DPM like your network's ✓ Improve Free Cash Flow by as much as 10% of the gross revenue	●	• Establish Billing DPM as a measure for defect reduction: for contracting, ordering, provisioning, billing, and maintenance processes • Implement Co1, Co0, and CoNone to reduce Billing DPM from the high of 100,000 to 100's	● ●
Minimize cost of SLA compliance ✓ Improve Free Cash Flow by as much as 5% of the gross revenue	●	• Establish Best-In-Industry QoE	●

Figure 3.25. Overall PMO gap closure.

Need	PMO	Human Resource Solutions	FMO
World-Class Trained Personnel	●	• **Establish BII Practices:** • Co1, Co0, CoN, DBOR, SDR, Tplan, Qplan, WR, DMOQ Goal setting, Process Management	●
		• **Establish BII Operations Discipline:** • Closed loop dashboard • Customer Communication, AYS, NE, 3CP, RA, RR, OM, M&Ps	●
		• **Train personnel**	●
		• **Enforce adherence to the BII practices**	●

Figure 3.26. Overall PMO gap closure.

Our conclusion from many assessments that we have completed is best summarized as (a) BII QoE performance is achievable and within reach for all CSPs, (b) the blueprint for the transformation can be put in place in a matter of few months to address the totality of the issues as outlined above, (c) the key programs include Optimization, and Transformation, (d) Human resources are systematically trainable to support the transformation, and (e) the urgency and justification for the transformation can easily be supported by a fact-based competitive analysis of the PMO.

CHAPTER 4

FMO
The Network & Systems

4.1 FMO: Introduction

Developing the FMO is a daunting task for the CSPs. After all, it is about redefining the CSP, articulating the need to move on to a unified network and preparing to offer DSP services, the workforce training, and the financial implication of it all.

The CSPs must be focused on developing a sustainable vision, the specific targets to achieve results for the business through transformation, and the specific design objectives for each functional entity to focus the energies in the direction of achieving the business targets.

The FMO must be driven by a vision, to be used as a guiding star to achieve DSP status. The vision is a guiding star for the creation of the FMO for a digital lifestyle, and the journey to it. This vision is a critical input for the alignment of the daily activities of 1000's of process associates to ensure that the DSP's energy is directed all going forward.

The vision can be operationalized once it is guided by specific business targets, over time, to delight the customers, to empower the workforce, and to create value for the shareholders. It is the job of the CEO and the executive team to establish the business targets, and to cascade the business targets to the design objectives for the operating units. The design objectives are the extension of the business targets, and the key to ensuring that all business processes are aligned with their transformation, and focused on their journey to the FMO.

In this chapter, we describe the FMO Network. In this process, we describe what must be done to establish the Transformation Targets, then convert those targets into "Design Objectives" to guide the transformation activities, then leverage the "PMO Gap Closures" to develop the FMO for bridging the gap from CSP to DSP.

4.2 FMO: Vision & Targets

Digital Lifestyle is the force to shape the future of the telecom. This future of telecom must be seen in the cross-section of four drivers: devices, content/applications, wireless, and wireline networks. This cross-section is evolving, and based on our collective insight by 2030, it brings about the ambient solutions.

Ambient solution for consumers is where the IoT sensors are sensitive and responsive to the presence of people; and for the business, it is about collaboration, sound, fact-based decision making, and availability to work from anywhere at any time with high availability services.

For the devices, it will go to common robotics, fueled by IoT sensors and Robotic vision.

For the content and applications, it will take us from basic subscription services (such as interactive video, music, magazine and news) to tele-immersion and ambient intelligent solutions fueled by Cognitive SMEs.

For mobility, it is the 7G (10 Gbps) with 100-to-1 wireless vs Wireline endpoints.

For wireline, it is the Exabyte backbone and Yottabyte data centers fueled by the unification of the wireless and wireline networks, 100 Gbps last-mile, cloaked network, and on-demand reconfigurable networks, where a network is being configured in real-time to deal with catastrophic events, e.g., moving traffic from the harm's way.

At this cross-section, the people are in the center stage for working productively, and having fun at it too. The devices enable big data and intelligence gathering; the wireline enables the exchange of big data (east-west pipes in addition to the north-south) over the geographical environment; wireless enables mobility and anyplace communication at High-speed of Gbps+; the applications and content provide insightful context to the big data which will facilitate human decision-making and entertainment.

The vision for CSP transformation must place emphasis on connecting people globally, providing meaningful content and applications, enabling decision-making, and enabling people to work and play seamlessly anywhere, with any device, and at any time.

The vision, constructed from the above context, will remain unchanged over time, but is achievable through a series of transformations, driven by putting the needs of the customers first, deployment of timeless BII practices and enforcement of operations discipline, vertical integration with content providers, deployment of the on-going changes in new technologies, and adapting the ever-changing governance/ regulatory requirements.

This vision must be driving the DSPs to promptly offer both legacy and futuristic services, such as multimedia and data, to ensure that the same set of services and user experiences are available over any existing and future access technologies, and to deliver a seamless handover of the live session between any pair of supported access technologies or devices.

The roadmap to operationalize the above vision will have several milestones. The first milestone over the next 3–5 years must focus on BII QoE, Network unification leveraging off-the-shelf HW/SW, and time-to-market Service Introduction platform for SoIP.

Given our executive and hands-on experience with major transformation projects at the Tier 1 CSPs, we strongly recommend aggressive transformation targets for the first milestone:

- Connectivity and digital services:
  - Revenue:
    - ✓ Up to 20% increase every other year due to new Digital Services
    - ✓ Up to 10% increase every other year due to footprint expansion
  - Performance:
    - ✓ Bandwidth in 5 years:
      - Consumer: 100++ Mbps symmetric, with bursting to Gbps
      - Business: On-demand > 40+ Gbps symmetric
    - ✓ QoE Driven Performance Targets (5000 Kilometers network):
      - Latency < 80 ms RTT
      - Jitter < 10 ms
      - Packet Loss < 0.01%
      - Service Availability (Biz) > 99.999%
- Capital unit cost reduction (ETE services for wireless/wireline/and video) > 70%
- OPEX Productivity target > 80%
  - ✓ Zero cycle time for customer-facing self-servicing processes
  - ✓ Zero defects for network-facing processes
- Scalable ETE Network:
  - ✓ Scale: 100's of Tbps per node
- Macro Service Catalog:
  - ✓ High Resolution (4K, 8K, 3D) Interactive Video, Streaming, Browsing, High-speed symmetric Access, Service Availability

4.2.1 Revenue/New Service Target

The revenue target is the key driver for the business, it sets the stage for CAPEX expenditure, new service introduction, and expansion of the service footprint to increase market share, etc.

Voice ARPU is dead; it is on a continuous decline at a rate of up to 50% every year. At the same time, the data ARPU is improving at the same rate as voice decline. However, the CSPs' data ARPU has mostly been focused on the increase in the bandwidth, not the actual service content. The service content has a tremendous upside potential, given the future OTT data services, such as OTT video, social services, gaming, and IoT driven services, and the incarnation of the current trends.

Given our collective experiences in new service introduction, it is a reasonable assumption that there will be an increase in total revenue of 3–5% YOY for a tier 1 DSP with the breakdown of up to ~ 1.5–2.5% for new services, as well as up to a 1.5–2.5% increase due to footprint expansion and market share.

4.2.2 Bandwidth Target

CSPs with quad-play connectivity service must establish their competitive bandwidth requirements for businesses and consumers. This bandwidth must be established for the delivery of a sustainable bandwidth bit rate target as well as a bursting target.

Given the emerging new content being offered at 4 k+ bit rate, a typical household with 2–3 TVs, home monitoring applications, as well as a couple of internet sessions could easily consume a constant bit rate of > 100 Mbps symmetric, bursting to 1+ Gbps. However, today's' competitive targets for consumer bandwidth is at 10's of Mbps asymmetric services. For business customers, the bandwidth requirements are based on customer demand above 10's Gbps symmetric services.

CSPs, across the world, have built their networks based on "Asymmetric Communication" technologies for consumers, and "Symmetric Communication" technologies for businesses. The fundamental assumption for consumers is that the downlink speed is significantly higher than the uplink speed. As a result, the spectrum allocation (e.g., in DSL, GPON, wireless) is designed and deployed with this mindset for consumers. This approach has forced CSPs to deploy two different sets of high unit cost-intensive platforms, one for consumers, and a separate one for businesses (requiring separate packet processing, fiber connectivity, call centers, etc.).

However, with the emergence of IoT sensors and the associated applications, such as real-time video surveillance, this legacy network mentality must switch to "Symmetric Communication" technologies in order to significantly improve the throughput in the last-mile access, to free up the bottleneck in the uplink speed, and to significantly lower the CAPEX unit cost for the businesses and consumers by consolidating the consumer and the business platforms into one. Today, the technology is there to make this convergence happen. It is a big deal with the significant unit cost reduction of the business platform, leveraging the emerging technology of business-grade platform (five 9's of availability) with 10+ Gbps EPON for the wireline last-mile access as well as cell site/cellular access point backhaul.

A network transformation must take into account the future applications that must leverage "Symmetric Communication" technologies, such as IoT sensors. IoT sensors are impacting access, aggregation, and the way the data centers are architected (leveraging Edge Data Center concept) and connected to support the applications that leverage the big data across a wide geographical location.

Given the state of bit rate for 4 K content and IoT services over the next several years, consumers will be demanding more and more bandwidth over 100's of Mbps symmetric bandwidth. We highly recommend symmetric bandwidth vs asymmetric at the rate of > 100's Mbps. IoT applications, such as video surveillance, will require symmetric bandwidth. CSPs' decision on the symmetric bandwidth offerings is critical to the investment required in the content platforms, access, aggregation, and the core network.

4.2.3 QoE Driven Network Performance Target

We strongly recommend that the network must be built to deliver customer experience performance for the 7 key macro service flows. These QoE performance targets (Dai,

2011) are key in providing seamless services to the end customers. Major services for the FMO are delivered through seven key service flows:

Macro Service Flows	QoE Expectation
Browsing- Service flow	• Near real time page download
VoIP- Service flow	• Voice quality Same as wireline • Call set up cycle time same as wireline
Streaming- Service flow	• Zero pixilation • Zero freeze • Sync'd voice & video
Interactive Video- Service flow	• Same as streaming • Action Trigger sync'd with video
Bandwidth- Consumer	• Guaranteed throughput (DL/UL)
Bandwidth- Business	• Guaranteed throughput (DL/UL)
Service Availability (Biz.)	• SLA compliant

Figure 4.1. QoE for the 7 macro service flows.

These QoE expectations must be cascaded throughout the network in terms of KPIs for response time, latency, jitter, packet loss, bandwidth, and service availability.

Once the network is transformed and ready for the delivery of these 7 flows under well-defined and engineered failures during the peak traffic time, and the additional bandwidth demand is driven by the bit-rate of the content (e.g., for each channel of HDMI @ ~ 6 Mbps, 4K at ~ 15 Mbps, and data @ > 100's Mbps through multi Gbps, and future bit rates dictated by the future content such as 3D, holographic, etc.), then any of the new and emerging telecom services can be delivered flawlessly over the ETE network.

ETE network must be designed for the most demanding service flow QoE, and it must be built with a scalable capacity in order to meet the growing demand of the end customers. The digital service providers (content, video, data, wireless, and VoIP) are to engineer their network for Interactive Video service flow as well as the last-mile bandwidth for consumers and businesses, where the Triple-Play connectivity service providers (Data, wireless, and VoIP) may do so for Streaming/VoIP service flows. In all cases, the service availability must be provided at five 9's of availability.

The service flow for interactive video must be used as a base minimum bandwidth to handle service flows for 4K TV, and future High-speed content.

FMO must focus on two sets of services: (1) Macro Service Flows-which is the building block of all DSP services, for which the network must be designed and engineered to deliver these service flows; and (2) Application-Specific services that leverage the network designed for the Macro Service flows, to deliver application-specific services consistent with QoE expectations and SLA, e.g., in-car emergency services, connected car, home automation, HSIA, etc.

A typical DSP network must be engineered for the "Macro Service Flows". Once the network is designed and deployed, meeting the stringent performance requirements of these Macro service flows, then any application-specific services can be introduced and delivered on that network with the appropriate QoE performance.

Given our expertise in network transformation, the Macro Service Flows have a massive implication from a network architectural perspective, for being resilient in case of failures, as well as requiring a well-orchestrated closed-loop real-time performance management system with the drill-down capability to root-cause the outage/degradation in QoE/SLA, and to enable quick actions when needed.

At the same time, the Application Specific Services presents a broad base implication on the SoIP platform, OSSs and BSSs, such as contracting, ordering, provisioning, maintenance, billing, etc., for service delivery, billing, and assurance.

FMO's seven Macro service flows will cover all services, today's and future quad-play services for Wireless, Voice, Data, and Video. The table below (Figure 4.2) presents the QoE Expectation, as well as the QoE targets for the 7

Macro Services	Acceptable	Unacceptable
Browsing- Service flow	Page Download: < **3** Sec	Page Download: > 8 Sec
VoIP- Service flow	Latency: < **170** ms Jitter: < **10 ms** Packet Loss < **0.01%**	Latency: > **250** ms Jitter: > **30 ms** Packet Loss > **0.04%**
Streaming- Service flow	Latency < **100** ms Jitter < **10** ms Packet Loss < 0.01%	Latency: > **120** ms Jitter: > **30** ms Packet Loss > **0.04%**
Interactive Video- Service flow	Latency < **80** Jitter < **10 ms** Packet Loss < **0.01%**	Latency: > **100** ms Jitter: > **30 ms** Packet Loss > **0.04%**
Bandwidth- Consumer	Guaranteed: Service flow bitrate /by market Bursting: 1G+ bps	**No Guaranteed bandwidth**
Bandwidth- Business	Guaranteed: symmetric, up to 40Gbps	**< Biz demand**
Service Availability (Biz.)	> **99.999** %	**<** 99.99 %

Figure 4.2. ETE QoE performance targets.

Service Flows. It provides what we consider as acceptable (Green) and unacceptable (Red) performance targets for the Service Flows in terms of response time, latency, jitter, packet loss, bandwidth, and availability. These targets are confirmed by the standard literature and are modified to reflect the live network tests, as well as end-user's expectation and feedback at major global tier 1 CSPs.

Internet browsing QoE target is established to ensure the user experience is met for the opening of a typical web page of ~ 9 MB in near realtime. This target of < 3 seconds is what the user's experience should be set at, for all access technologies including wireless, xDSL, and xPON. We also established that the wait time of > 8 seconds is not acceptable. To meet the demand of this target, it maybe required to have minimum bandwidth requirements on the last-mile connectivity infrastructure; the proximity of the content (at the edge or in the aggregation network) to the eyeballs is also essential in ensuring this QoE performance.

VoIP QoE target (Jelassi, 2012) for < 170 ms one-way latency is critical to ensure equivalent to circuit switch quality of experience, also, call set up cycle time must be close to the wireline experience (e.g., << 3 seconds). Jitter of more than 10 ms (ETE) presents voice clipping and packet loss of > 0.01% presents excessive load on the network, which is the root cause for voice clipping.

Streaming service flow (Barman and Martini, 2019) is sensitive to latency over 100 ms as it causes video freezing, pixellating, and often out-of-sync audio and video when audio is leading video by more than 80 ms or lagging 160 ms behind the video.

For most of the video streaming protocols, TCP and UDP serve as the underlying protocols. UDP is preferred over TCP as TCP introduces delays due to retransmission to handle the packet loss due to network congestion, and is reflected in customer experience as video freeze. UDP, on the other hand, does not care about the packet loss, and as a result it could easily cause picture freezing and pixelation.

Jitter of more than 10 ms presents audio and video clipping and packet loss of > 0.01% puts an excessive load on the network, which is the root cause for audio clipping and video freeze.

Interactive video (IV) (Barman and Martini, 2019), A cloud gaming system must collect a player's actions, transmit them to the cloud server, process the action, render the results, encode/compress the resulting changes to the game-world, and stream the video (game scenes) back to the player, e.g., gaming application. This process presents a serious challenge when the latency exceeds 80 ms, specifically for a gaming application that requires low latency, such as First-Shooter-Experience (FPS). It has all the similarities with the video streaming service flow, and more for making the gaming experience acceptable.

IV service flow is sensitive to latency over 80 ms for gaming applications (e.g., FPS) as it causes the "action trigger" to be out of sync with video rendering.

Bandwidth-consumer, it is aggressively moving to symmetric bandwidth given the emerging applications, such as video surveillance and IoT. A typical home with 5–10 video cameras operating at HD resolution could easily consume > 25 Mbps uplink. When you add to these other home-data applications, it is more than 80 Mbps uplink and much more for the downlink, given the 4K+ content and other data services.

Bandwidth- business, it is aggressively moving to 40+ Gbps symmetric bandwidth, given the reliance on the need for massive processing through data centers. A typical business (Small, Medium Enterprises) connectivity requires an excess of 10+ Gbps, and it is expanding rapidly to 40+ Gbps.

Service availability, the key business customers are demanding a five 9's of availability (Greene and Lancaster, 2006) as a necessity for mission-critical applications, and they are willing to pay for the premium. Five 9's of availability is a pretty tough target to hit. This corresponds to 5.26 minutes of unplanned downtime in a year.

To recover from an outage within a 5.26 minutes, it is rather unlikely that if the NOC engineers can attend to a restoration within 5.26 minutes, i.e., to receive the notification of the outage, auto process the alarm, display the alert information, corollate the alarm to other states in the network, have a NOC engineer acknowledge the alert, diagnose the root cause, determine the likely fixes, open a trouble ticket, dispatch technicians, fix the root cause, configure the device, launch the device activation tools, and restart the service.

This lengthy process is demanding that the entire DSP platform must be resilient and must withstand failures if it is to deliver five 9's of reliability.

The resilient network architecture must be established based on the Co0 for zero-demand maintenance in the FMO, to deliver at this performance level and better.

4.2.4 CAPEX Target

The CAPEX ratio for the BII CSPs (CAPEX divided by revenue) must be maintained between 12% and 18% (A future for mobile operators: The keys to successful reinvention, 2017) to ensure that the CSP is properly investing in the new technologies and growth. CAPEX ratio of > 18% is indicative of a variety of reasons, including adding capacity with higher unit cost, adding too much capacity ahead of the customer demand, or a network that was neglected for proper investment on an annual basis. CAPEX ratio of < 12% is also indicative of under-investment in new technologies and upgrades.

It is often the case when ARPU is growing at a much lower rate than the growth in network traffic, then the response to the traffic demand leveraging the existing network gears will force the CAPEX ratio to exceed the upper bound of 18%. Actions must be taken to lower the CAPEX unit cost on the growth, leveraging architectural transformation, as well as new and improved network technologies.

Given that video content is dominating the ever-increasing network traffic, and almost the flat ARPU, it is a kiss of death if the CSPs continue to grow the network with the legacy technologies, processes, and OSSs/BSSs.

We need to focus on the CAPEX unit cost reduction of the end-to-end cost drivers to focus on transformation where it is most needed. The transformation must be guided by what the cost drivers are, and what the performance gaps are. The cost drivers must be established for ETE connectivity services:

A. wireline services, measured by GigE equivalent connectivity for ETE services;
B. wireless services, measured by Million minutes of equivalent usage (dominated by video); and
C. video/Content services, measured by unit cost for the living units passed.

Once we determine the CAPEX unit cost for each ETE service, and establish the targets for CAPEX unit cost reduction by each service, given its projected traffic growth and ARPU expectations, then it is easy to cascade those targets to major components of the ETE network that is impacting the services.

The major components of the cost of ETE network are found in the following key domains of the cost structure: for GigE wireline Customer services, it is the CAPEX unit cost in the backbone, aggregation, access, and the data centers; for wireless customers, it is the unit cost in the spectrum, backbone, packet and circuit switching infrastructure, the cell site backhaul, the RAN (Radio Access Network), and the data centers; and for the video customers, it is the unit cost for the backbone, video/content processing, and storage, switching, and the last-mile connectivity with FTTx.

Next we explain an example of what the CAPEX targets could be set at for transformation.

Overall CSPs must introduce new network technologies to ensure a minimum 70% reduction in the CAPEX unit cost over a 2–5-year period to be able to maintain competitive pricing for its growing traffic and services.

We foresee a traffic growth of 50% YOY in the core backbone, and 30% YOY on the access side of the CSP network. The difference in the growth is dependent on the via traffic that passes through the core backbone. This growth in traffic must be delivered with the CAPEX unit cost consistent with the increases in the ARPU.

For example, in many cases, given a flat ARPU in the business plan, and a small increase of ~ 3% in incremental revenue YOY, and an increase in the traffic of 30% in the access and 50% in the core YOY, we calculated and established (for all traffic growth) a 50% CAPEX unit cost reduction for the wireline GigE equivalent services, a 70% reduction for wireless, and 55% for the video services.

Next it was key to cascade this reduction target to the components of the network. In that, for wireline services (with 50% reduction in CAPEX unit cost) we needed a 65% reduction in the backbone, a 40% reduction in the aggregation, and an 80% reduction in the last-mile access part of the network.

The same analysis dictated that, for wireless services, we needed a 70% reduction in the packet and circuit switching infrastructure as well as cell-site backhaul, and a 75% reduction in the RAN unit cost.

And finally, for the video/content services, we needed a 25% reduction in the video/content processing, storage and switching, as well as a 75% reduction in the FTTN and home wiring.

With this approach to target setting, BII CSPs are ready to unleash their innovative talents to develop the overall architecture, and to identify the specific solutions needed in every part of their ETE networks. Generally speaking, we achieved the target reductions through: (1) convergence of the backbones and aggregation across wireline, wireless, and video services (FMC); (2) convergence of the aggregation network and CSBH to an all IP infrastructure; (3) convergence of the wireless packet and circuit switching platforms; (4) SDN; (5) unified Data Centers by migrating to NFV; and (6) eliminating the need for in-home wiring by converting to the wireless distribution of video in the home/premise.

4.2.5 OPEX Target

The OPEX productivity target must focus on defects and cycle time reduction throughout the processes inclusive of OSSs and BSSs. This target must be set: (a) to match the network DPM for its' business processes, (b) to drive the cycle time to zero for all customer-facing business processes to enable self-servicing, and (c) to reduce the cycle time for network-facing processes significantly in order to ensure the resiliency of the network and its services are not compromised due to the cycle time for time-to-repair.

As the ETE network is transformed to deliver packet loss at less than 50 defects per million (DPM), the DSPs' processes, and systems must be driven by the same performance targets as the network DPM. This target is the key to significantly improving OPEX performance.

Given our experience, the DSP billing contra revenue relative to the total revenue is an excellent source to use, to quantify the defects in the overall DSP PMO processes and systems. We strongly recommend (Billing DPM) to drive the ETE transformation in the overall CSP processes inclusive of marketing, contracting, ordering, provisioning, test and turn up, customer care, maintenance, and billing. The big picture of a CSP overall financial status requires that CSP establish a reserve, called contra revenue, to compensate for the lost revenue due to errors in their bills for the customers. We define "billing DPM" as a ratio of "Correct Charge Adjustment" to the "Total Billed Revenue", normalized to a million.

For example, a major Tier 1 CSP had their billing DPM at 100,000 before the transformation. In this case, we establish target "Billing DPM" of 3000 in 3 years, and 100 in 5. The financial impact over 3 years was 30X reduction in the contra revenue, which directly impacted the free cash flow.

The following actions are the driving force for the reduction of the contra revenue to the tune of 30x reduction:

(a) OPEX unit cost reduction due to the convergence of systems with the concept of 1, e.g., one SoIP platform, one billing system, one ordering systems, etc.

(b) The resiliency of the network with the concept of zero customer-impacting defects, e.g., elimination of on-demand maintenance where network failures are no longer impacting the service; and

(c) Extreme automation for zero cycle time, with the concept-of-none hand-offs to enable self-servicing.

4.3 FMO Typical Services

A DSP with digital services as well as quad-play connectivity services should have a scorecard, inclusive of the following (Figure 4.3):

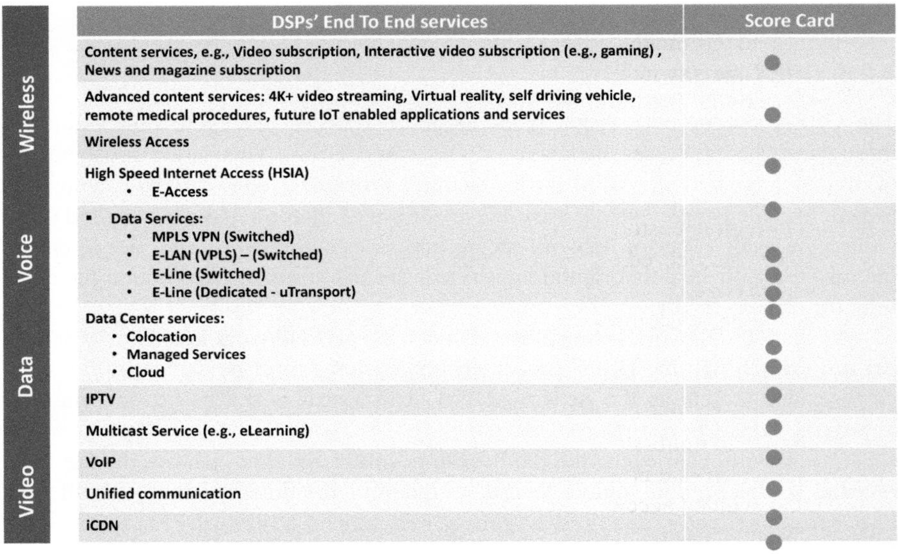

Wireless | Voice | Data | Video

DSPs' End To End services	Score Card
Content services, e.g., Video subscription, Interactive video subscription (e.g., gaming), News and magazine subscription	●
Advanced content services: 4K+ video streaming, Virtual reality, self driving vehicle, remote medical procedures, future IoT enabled applications and services	●
Wireless Access	
High Speed Internet Access (HSIA) • E-Access	●
▪ Data Services: • MPLS VPN (Switched) • E-LAN (VPLS) – (Switched) • E-Line (Switched) • E-Line (Dedicated - uTransport)	● ● ● ●
Data Center services: • Colocation • Managed Services • Cloud	● ● ●
IPTV	●
Multicast Service (e.g., eLearning)	●
VoIP	●
Unified communication	●
iCDN	●

Figure 4.3. FMO typical services

We do look at the DSPs as a provider of many services, including disruptive content services such as Netflix disrupting the way we watch movies, gaming services, magazine and news services, etc.; Wireless High-speed access; Wireline High-speed Internet Access (HSIA); Wireline Data Services, including MPLS VPN, Switched E-LAN (VPLS), Switched E-line, Dedicated E-Line; Data Center for colocation and Managed Services; IPTV; Multicast Service (e.g., eLearning); VoIP; Unified Communication; iCDN; Transit services; Cybersecurity services.

These services must be delivered over a single network infrastructure that is capable of handling the seven service flows, as outlined above.

Digital services are provided through a SoIP platform with the run-time-execution environment for all devices.

Consumer HSIA services are provided for consumers. This service rides on an IXP/GW infrastructure, through the backbone, the regional core and provider edge, service aggregation network, and the multi-service edge. The bandwidth will be driven by the customer's demand, in the excess of 100's of Mbps (with optional symmetric bandwidth).

The E-Access services are primarily provided for the Enterprise customers. This service rides on an IXP/GW infrastructure, through the backbone, the regional core and Provider Edge, Service aggregation network, and the multi-service edge. The bandwidth will be driven by the customer's demand, in the excess of 10's of Gbps.

The MPLS VPN (switched), and E-LAN (VPLS switched) provide multipoint-to-multipoint (any-to-any) connectivity. E-LAN services are designed for multipoint Ethernet VPNs and native Ethernet transparent LAN services, E-LAN and MPLS VPN services are provided for the enterprise customers. This service originates from the enterprise customer premise, through the access/aggregation/regional core, backbone, and terminates on a similar path to the consumers and business customers.

The E-Line (switched) and E-line (dedicated transport) provides point-to-point connectivity. E-Line services are used to create Ethernet private-line services, Ethernet-based Internet access services, and point-to-point Ethernet virtual private networks (VPN) services. E-Line is provided in the region and across the region to the business and consumer customers.

For in-region customers, the service will be connecting customers from an ingress multi-service edge, through the aggregation, and terminates at the egress multi-service edge.

For national customers, the service will be connecting customers through the backbone as well.

The IPTV services are provided in the region, and it is originating from a data center connected to the edge of the network, through the regional network, aggregation, and multi-service access, terminating at the customer premise.

Multicast service is provided at the regional and national levels. It is provided through the enabled access, aggregation, and backbone network.

iCDN services are provided both off-net, and on-net.

The off-net iCDN service is provided through the internet, IXP/GW infrastructure, and the data centers (iCDN nodes) connected to the nearest PE of the cooperating ISP networks.

The on-net iCDN service is provided through local content residing in data centers attached to the PEs, and through the aggregation network, and the multi-service access. In cases where the data center is not attached to the serving PE, the service will be provided through the backbone network connected to a remote PE.

4.4 FMO: Assessment of Available Network Technology Performance

The assessment of the network technologies is aimed at determining if a legacy network, or a new one, built with the today's and emerging technologies can withstand the stringent requirements of QoE for the evolving Macro Service Flows. We have assessed the key technologies in the access, aggregation, core routing, content & storage, services complexes.

Based on our analysis, we have established,once the FMO network is properly transformed leveraging today's and emerging technologies, the QoE performance for the critical flows will be delivered, with some exceptions, specifically xDSL and

TDM access. Also, our analysis indicated that some emerging technologies, such as High-speed 10G+ EPON, offer significant advantages over legacy PON and legacy CWDM-based technology (e.g., SONET/SDH, OCx) for a scalable, "one-platform-access" for both business and consumers.

The transformed network must be supported by areal-time performance management system with drill-down capability as well as analytics to enable root cause analysis and actions. The ETE service performance management requires QoS performance management at every component of the ETE network. It also requires placement and activation of data gathering probes throughout the network, and establishment of real-time dashboards to enable surveillance, analysis, notification, and actions to ensure sustainable service performance. We have concluded that today's technologies in the access through core and content are capable of enabling real-time monitoring, analysis, self-healing and alerting.

The technology assessment is essential to authorize the CSP with an "Approval for Use". It starts with a request for new services/products and technologies at a transformational unit cost, triggers the development of network technology as well as the development of IT in support of the new introduction, and ends with the preparation of the personnel, the service, and the network for what is called FOA (first office application). The approval for use requires a mirror image of the production network as a testing platform. It must include testing of system functionality, security, performance, scalability, physical design, and documentation, before issuing the "Approval for Use". Generally speaking, given the scale of tier 1 CSPs, and the scope of use, we have determined that the network gear manufacturers' specification for performance is usually off by ~ 20%. for example, if the manufacturer is stating a safe operating performance of 95% utilization, say for the CPU on their network gears, you must assume a performance level below 80% before you have packet drop, and degradation in QoE performance.

FMO's ETE network components include access, aggregation, core, cybersecurity (included in the content) and content, as shown below for wireline and wireless. The rigorous testing process that we followed presented us with the following results:

The performance of access to the content is highly dependent on the types of last-mile access technology. The aggregation, core, and the access to the content can be relatively and easily transformed to deliver predictable ETE performance. However, when taking into account the last mile connectivity, the QoE performance is widely different. Here is what you should be able to expect from a properly transformed ETE network:

In the example below (Figure 4.4), we have assumed a national network with coverage of up to 5000 kilometers. As you can see, the most demanding service flow (Interactive Video) can be supported in a transformed network, provided the architectural option is considered to place the content close enough to the eyeballs.

The key issue is the last mile connectivity of xDSL with which the interleaving issue puts a significant limitation on network latency. It is possible to minimize the interleaving issue by placing the content much closer to the eyeballs, however, to provide IV services, this will significantly limit the geographical coverage for such services. Furthermore, the bandwidth limitation of the copper infrastructure is not acceptable given the consumer and business demand for a much higher bandwidth.

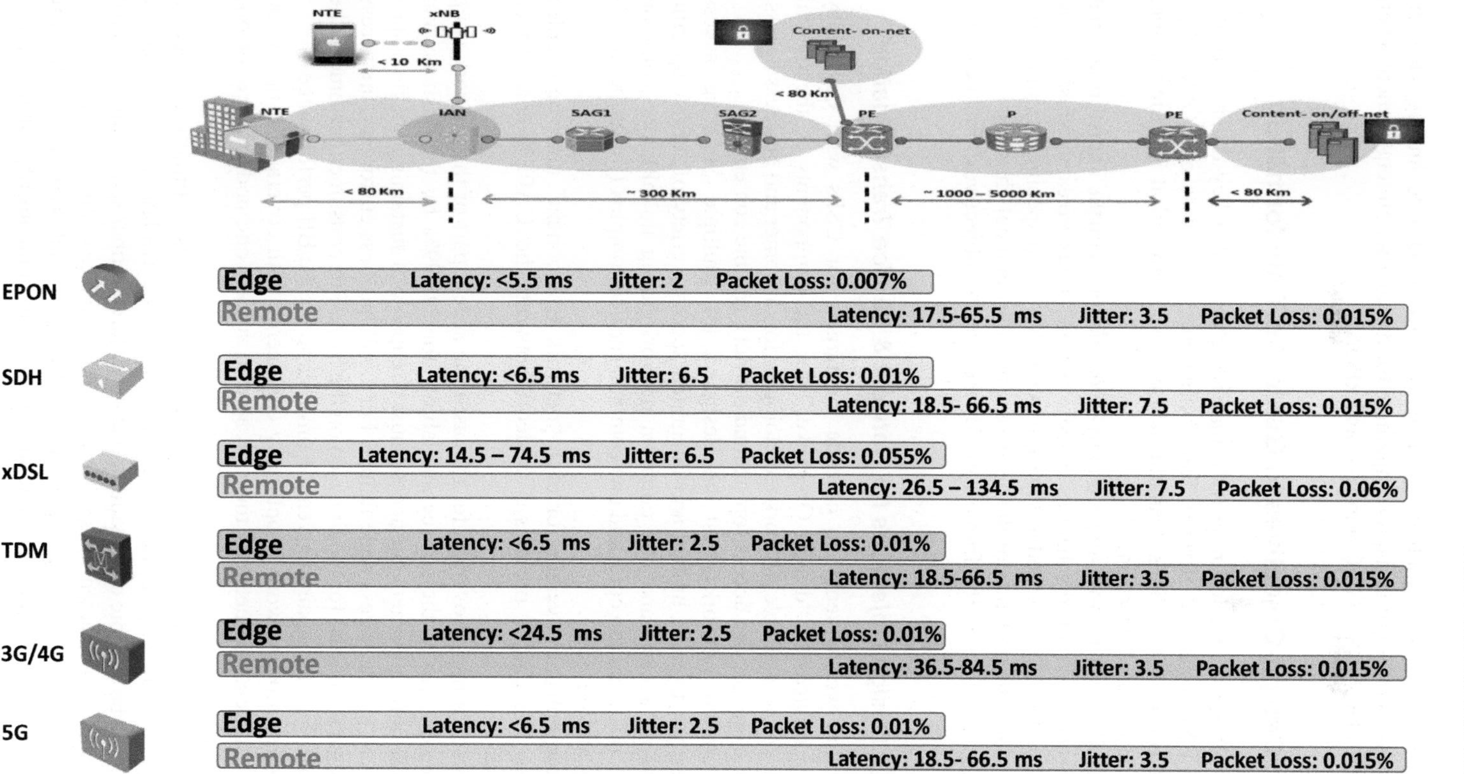

Figure 4.4. QoE performance expectation-EDGE vs remote data centers.

We strongly suggest to gradually migrate away from xDSL technology. It is an obsolete access technology, and not competitive for the bandwidth requirement when compared with DOCSIS, 4G/5G wireless, and xPON.

4.5 FMO: Design-Objectives to Guide the Transformation

A typical CSP transformation requires a massive change in 5 key domains of the business: (1) Network and Service Transformation, (2) Process and OSSs/BSSs Transformation, (3) Operations Transformation, (4) Organization Transformation, and (5) Cultural Transformation.

It is crucial to align the massive company-wide changes with the Transformational business targets. This requires that the Transformation targets are established and are carefully cascaded and allocated as "Design Objective" for a specific domain in the business to guide the direction of the necessary transformational changes.

Below, we will cover the framework for the "design objectives" for these five domains.

4.5.1 FMO: Design Objectives for Network & Service Transformation

In this section, we'll attempt to cascade the targets for CSPs with Quad-Play connectivity services to "design Objectives" for the components of the overall network, given the available network technologies for customer and business access, aggregation, core, service access (on-net and off-net), and the content.

FMO unified network must be designed for multiple last-mile access technologies, yet for one, unified and integrated access (variety of service-specific customer-facing access, and one common network-facing uplink), one common aggregation, one common core, and one common service complex (for on-net & off-net contents and services).

The geographical coverage for each "network Component" must be decided carefully, as this choice presents a tradeoff between the CAPEX and service performance.

The most optimum coverage for the last-mile access part of the network is up to 80 kilometers. Given the service availability for 99.999% for the business and enterprise/medium customers, the access must be capable of sustaining one fiber span failure, as well as one access node failure. For small business customers, the access failure must be supported for repair based on the SLA. For consumer customers, the access failure must be repaired given the time frame for the BII performance.

The aggregation network is optimally engineered with coverage of up to 300 kilometers. Aggregation network must be able to sustain one node failure, or one fiber span failure.

The last mile connectivity to the service complexes is optimally engineered for coverage of up to 80 kilometers. To meet the service availability target, service complexes must be capable of sustaining one PE failure, or fiber span failure. Finally, the content must be protected for a minimum of two service complex failures.

The Core Backbone is engineered from 1000 up to 5000 kilometers. The core must be capable of sustaining one node failure, or 2 simultaneous fiber span failures.

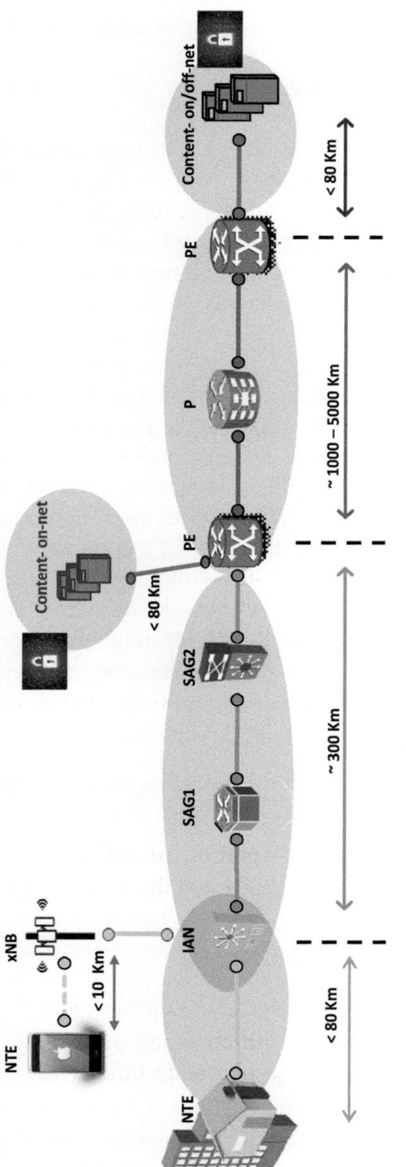

Figure 4.5. National network coverage.

The picture above (Figure 4.5) provides a simplified view of what we call an optimum coverage area for a national network. Given our expertise and extensive modeling and hands-on experience, this national coverage provides the most efficient balance between the CAPEX unit cost and the QoE (Just a quick reminder for the acronym—IAN: Integrated Access Node; SAG 1&2: IP based service aggregation, PE: provider edge, P: provider core router).

Figure 4.4 presents the ETE network performance guideline for QoE, given the national coverage for the ETE connectivity services.

As a reminder, the QoE for interactive video (ETE service) has an upper limit of 80 ms for latency (RTT), jitter of < 10 ms, and packet loss of < 0.01%. Once a CSP network is transformed, it must deliver this level of performance, or better, across the unified and the dedicated access infrastructure depending on where the content is located relative to the eyeballs, i.e., at a data center attached to a remote PE, at the data center attached to the homed PE, or at a data center attached to the aggregation node.

Figure 4.4 outlines the ETE network performance to be expected from a transformed network given the variety of the last-mile access technologies, as well as the placement of the content at the Edge vs Remote to the eyeballs. The underlying assumptions for this ETE performance are (a) national coverage area for the ETE network as shown in the picture, (b) RTT for optical ETE connectivity services at 10 ms per 1000 Km, except for the last mile for TDM, xDSL, and wireless, and (c) a 0.1 ms delay caused by each piece of equipment on the path of the ETE connectivity, which equates to 2 ms per 1000 Km.

As you see below, the weakest chain is the last-mile access, and it is the most capital intensive to transform. Therefore, a CSP must start by establishing the design objective for access as a priority, given the embedded base of their access technology.

The Last Mile Access part of the network performance is highly dependent on the underlying technologies, and its coverage area is up to 80 kilometers:

a. xPON: latency < 1 ms, Jitter < 0.5 ms, and packet loss < 0.001%, and coverage up to 80 kilometers.
b. SDH/Sonet: latency < 1 ms, Jitter < 0.5 ms, and packet loss < 0.001%
c. xDSL/EoCU: latency < 10–70 ms, Jitter < 5 ms, and packet loss < 0.05%, and the coverage up to 5000 feet.
d. TDM: latency < 2 ms, Jitter < 1 ms, and packet loss < 0.005%
e. Wireless: latency < 20.2 ms for 3G/4G and < 2 ms for 5G, Jitter < 1 ms, and packet loss < 0.005%, and the coverage up to 10 kilometers.

For any CSP where their access network is dominated by a twisted-pair copper infrastructure, it is highly recommended to allocate a latency target of maximum 50 ms for the access in order to ensure quad-play connectivity services are provided within the QoE performance target. Any homes or businesses that don't qualify and exceed the maximum latency must be flagged accordingly and not covered for quad-play connectivity services.

The core of the network must be transformed, making it capable of delivering latency (Round Trip) between 12 ms for a 1000-kilometer coverage under failure,

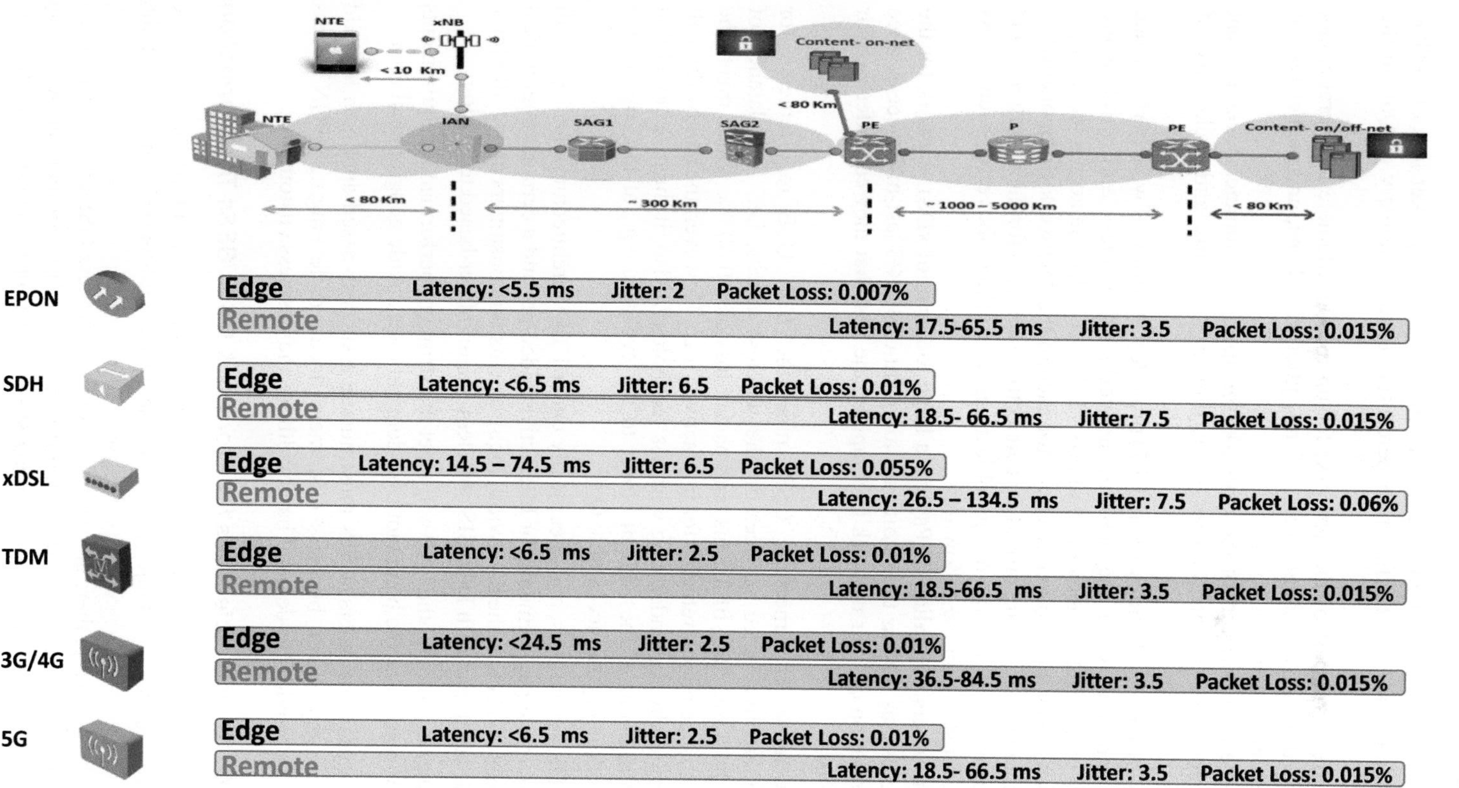

Figure 4.4. QoE performance expectation-EDGE vs remote data centers.

and up to 60 ms for a 5000 coverage, jitter of < 1 ms, and packet loss of .005%. This performance, at peak traffic of 95 percentile, ensures delivery of QoE in the backbone.

The aggregation part of the network, for up to 300 kilometers coverage, can be transformed to deliver latency of < 3.5 ms, jitter of < 1 ms, and packet loss of < 0.005% under failure.

The content Access part of the network, for up to 80-kilometer coverage, can be engineered to deliver latency of < 1 ms, jitter of < 0.5 ms, and packet loss of < 0.001% under failure.

The images below (Figures 4.6 and 4.7) provide a snapshot of what performance should be expected when the separate networks of wireless and wireline are transformed into a unified network, and provided over a diverse access technology. It also demonstrates the design objectives for each of the "Network Components".

The Global CSPs that are providing connectivity services for their consumer and business customers, given the design objectives as outlined above for each access technology, must expect the following:

For consumers: regarding CSPs with a large footprint of xDSL network for the last-mile access, as well as the significant footprint of 3G/4G last-mile access, once the transformation of their network is complete, the best that they can offer to their customers are:

(1) for wireless customers,the CSPs meet the ETE QoE requirements for the interactive video services on the wireless access, i.e., ETE latency of 24.5–84.5 ms, jitter of 3.5 ms, and packet loss of < 0.005%, which meets the most stringent requirements for interactive video. However, the bandwidth will be limited by the underlying wireless technology. Also, it must be noted that the placement of the content closer to the eyeballs, e.g., at the edge of the network is key for acceptable QoE.

(2) for xDSL customers, and specifically under FTTN architecture, the CSPs will not meet the requirements for Interactive Video beyond a certain geographical space, nor meeting the bandwidth requirements for consumers above 40 Mbps. xDSL not in par with DOCSIS, wireless 5G, and fiber-based technology.

(3) for fiber customers, the CSP will meet the stringent performance requirements for IV, as well as the bandwidth-on-demand over a wide geographical area.

For business customers: CSPs must pursue a success-based approach to build fiber connectivity to the business customers' location, leveraging xPON access technology to provide on-demand bandwidth for the business customers.

4.5.2 FMO: Design Objectives for Process & OSSs/BSSs Transformation

In this section, we'll attempt to cascade the targets for Quad-Play "service performance" to "design Objectives" for the Process/OSSs/BSSs/throughout the network.

Generally speaking, there are three categories of Process/OSSs/BSSs in a typical DSP: (1) Services over IP systems platform (SoIP), (2) Customer/partner-facing systems platform, and (3) Network-facing systems platform. A system platform in

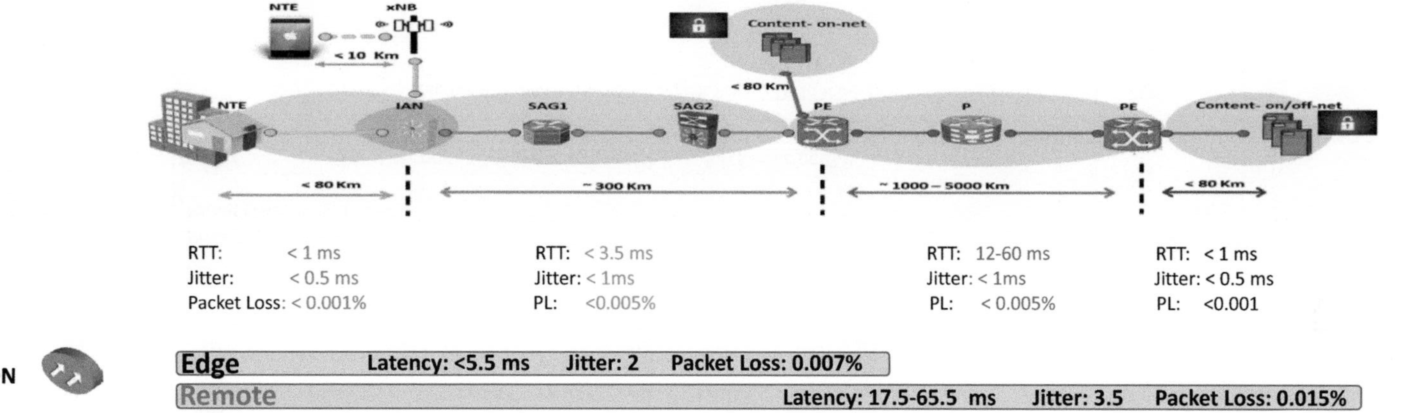

RTT: < 1 ms
Jitter: < 0.5 ms
Packet Loss: < 0.001%

RTT: < 3.5 ms
Jitter: < 1ms
PL: <0.005%

RTT: 12-60 ms
Jitter: < 1ms
PL: < 0.005%

RTT: < 1 ms
Jitter: < 0.5 ms
PL: <0.001

EPON

Edge Latency: <5.5 ms Jitter: 2 Packet Loss: 0.007%

Remote Latency: 17.5-65.5 ms Jitter: 3.5 Packet Loss: 0.015%

RTT: < 2 ms
Jitter: < 5 ms
Packet Loss: < 0.005 %

SDH 

Edge Latency: <6.5 ms Jitter: 6.5 Packet Loss: 0.01%

Remote Latency: 18.5- 66.5 ms Jitter: 7.5 Packet Loss: 0.015%

RTT: < 10-70 ms
Jitter: < 5 ms
Packet Loss: < 0.005 %

xDSL 

Edge Latency: 14.5 – 74.5 ms Jitter: 6.5 Packet Loss: 0.055%

Remote Latency: 26.5 – 134.5 ms Jitter: 7.5 Packet Loss: 0.06%

Figure 4.6. Network design objectives—EPON, SDH, xDSL.

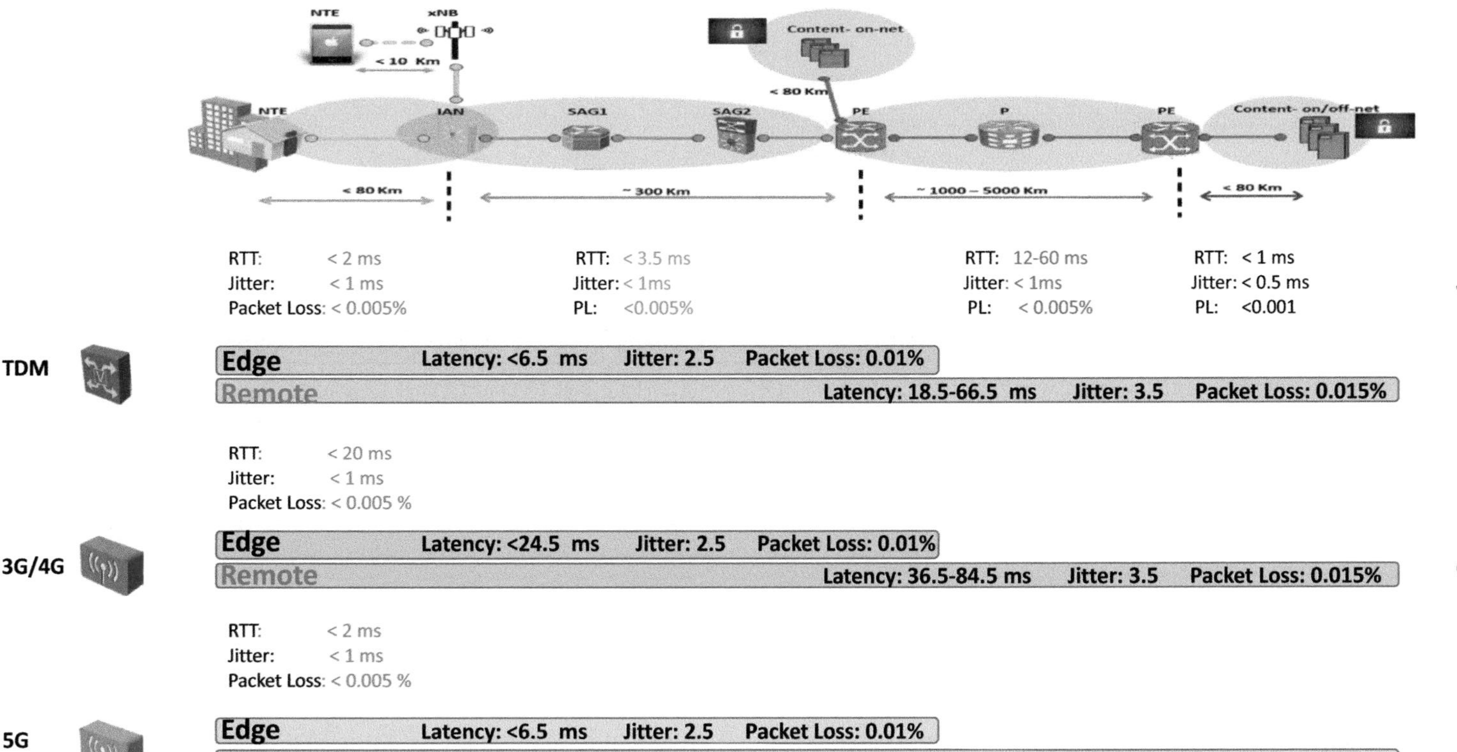

Figure 4.7. Network design objectives—TDM, 3G/4G, 5G.

this context refers to software/hardware/networking infrastructure upon which the platform is functioning. However, in this book, as we are not promoting a specific hardware, software or networking platform, we refer to a system platform by its functionality.

These three sets of processes and systems platforms must be built based on the Concept of One (for consolidated network, processes, organizations, and systems), Concept of Zero (for zero defects + zero cycle time), and Concept of None (for zero touch for customer-facing processes to enable self-servicing).

SoIP platform must be aimed at establishing an open platform for the fast introduction of new services, and to enable third party new service introduction.

The customer/partner facing processes (such as ordering, provisioning, billing, trouble management, service restoration, partner transaction) impact customer and partner satisfaction. These processes are critical for establishing a strategic advantage for the CSPs by being perceived as "easy to do business with". These processes must be driven to deliver in real-time with zero defects, and engineered to be self-servicing. Once these processes are re-engineered/transformed, you must expect significant improvement in the performance of the touchpoints, as well as a 40% OPEX unit cost reduction for a 50% cycle time reduction.

Your customer-facing processes and OSSs/BSSs must be transformed to enable self-servicing. This requires near-zero defects and near-zero cycle time. We strongly suggest that you focus your billing DPM to operate at near zero, E-bonding close to 100%. The table below (Figure 4.8) presents what the design objectives must be set at for the transformation of customer-facing processes and systems.

Customer Facing Processes:	FMO Design Objectives
E-Bonding (volume of transactions):	
• Ordering & Status	• >90% business orders
• Trouble Management	• >95% Trouble tickets
• Account & Billing info	• 100% transactions
• Inventory Management	• 100% transactions
E-Bonding (Web enabled Servicing):	
• Network Performance	• Through Portal
• Network Tools	• >95% Trouble tickets thru portal
Contra Revenue DPM	• <1000 DPM
Order Self-service (L-M-H Speed)	
• Business	• Real-Time (Point & click Through Portal)
• Consumers	• < 48 hours (calendar)
Fault Management	
• Service Restoration	• Real-Time
End-to-End performance monitoring and reporting (e.g., SLA)	
• SLA monitoring	• Real Time
• Fault and threshold reporting	• Real Time
• Performance reporting	• Daily/weekly

Figure 4.8. Design objectives for customer facing process/system transformation.

On the other hand, Network facing processes and systems (such as capacity planning, routine maintenance, etc.) are the support processes and systems, and are logistical in nature vs customer-facing. They must be driven by the urgency and efficiency to support customer-facing processes. You must expect significant improvement in the performance of the touchpoints, as well as a 40% OPEX unit cost reduction for a 50% cycle time reduction.

The table below (Figure 4.9) presents what the design objectives must be set at for the transformation of network-facing processes and systems.

Network Facing Processes:	FMO Design Objectives
Process Management	In Place
Capacity Planning & Delivery • Augmentation • Construction	 • Same day • 1-6 weeks
Order Self-service (L-M-H Speed) • Business • Consumers	 • Real-Time (Point & click Through Portal) • < 48 hours (calendar)
Fault Management • SPOF • Restore / Repair • Mean Time To Repair (MTTR) • Time to Restore (TTR) transaction volume	 • 0% SPOF • 100% service restored first, and repaired later • By Class of severity • >90% TTR < 2 hours
End-to-End performance monitoring and reporting • QoE monitoring and reporting • Domain performance reporting	 • Real Time • Real Time
Inventory Management • Plug-n-play • Auto Inventory • Auto supply chain	 • Real Time • Real Time • Real Time

Figure 4.9. Design objectives for network facing process/system transformation.

4.5.3 FMO: Design Objectives for Network Operations Transformation

The FMO network architecture, as will be described later, must be built for survivability, and it is the daunting task of the network operations function to maintain the overall network in a survivable state at all times.

The network operations' design objective (Figures 4.10 and 4.11) must be set to ensure flawless execution of network operations and ensure the five 9's of service availability for the customers at all times.

Operations Discipline:	FMO Design Objectives
Network Performance • Rule based (Capacity management)	• Manage to QoE and SLA targets • Real time drill down capability to root cause • 100% rules driven
Customer Care / Communication	• Morning report / weekly Report thru the portal
Restore First / Repair next • MTTR (to repair) • MTTR (to restore)	 • Within 2 hrs • Real time
TTR Distribution	• >90% TTR < 2 hrs
Outage Management • MCB/TCB	 • 100% compliant with Restore/Repair 100% service verification after Restore/Repair
Ask YourSelf Compliance	100% NOC and Technician compliant
Methods & Procedures • % network activities supported by M&P • % compliance with M&P	 • 100% • 100% compliance

Figure 4.10. Design objectives for operations transformation.

Operations Discipline:	FMO Design Objectives
Network Events	• 100% monitoring when performing activities • Zero POE
Root Access	• Restricted • Known impact
End-to-End performance monitoring and reporting • QoE monitoring and reporting • Domain performance reporting	• Real Time • Real Time
Large scale change (migration/upgrades)	• Planned restoration • No outage
Performance & system health management	• Real Time system data collection • Real time ability to alert on threshold violation
Traffic data analysis and data mining • QoE monitoring	• Full traffic breakdown by services • Ability to identify bandwidth abusers
New service / product introduction	• 100% completed with NVT/ORT
Disaster Recovery	• Full recovery within 72 hours

Figure 4.11. Design objectives for operations transformation—continued.

4.5.4 FMO: Design Objectives for Organization Transformation

FMO organizational structure must minimize vertical, silo-based, multi-layer, decision-making and multi-matrixed reporting relationships. The process associates in the FMO network must be held accountable for the impact that their actions are having on the end customers. They must be enabled to make decisions, under business rules, without any delays imposed by the management hierarchy. Figure 4.12 presents a set of organizational targets for transformation.

Organizational Attributes	FMO Design Objectives
Tier 1 CSP Management Layers	• < 7
Accountability	• Focused on the ETE Process • DMOQ focused • Reduced hand-offs • Focus on cycle time • Supported by KPIs
Decision Making	• Decentralized for Day-to-day decisions & supported by rules and KPIs • Centralized for structural changes

Performance goal setting process	• DMOQ set by senior leadership and CFO
	  • cascaded to BU business goals
	  • Cascaded to BU performance targets
	  • Cascaded to BU ETE process KPI targets
	  • Mapped to specific funded initiatives

Benefit realization program management function	• In place until one year after completion of transformation

Figure 4.12. Design objectives for organizational transformation.

4.5.5 FMO: Design Objectives for Cultural Transformation

FMO cultural transformation must focus on changing the behaviors of individual process associates throughout the CSP, which requires the learning of new and transformational behaviors.

The transformational behavior must value rewards for impacting QoE, process associates who are operating in the process and consistency of the output of the process for impacting QoE, accountability for impacting QoE, urgency for impacting QoE, and fast decision making for impacting QoE. Figure 4.13 presents a set of design objectives for cultural transformation.

Cultural Attributes	FMO Design Objectives
DMOQ driven behavior	• All employees are part of a one or more ETE processes • Payout driven by ETE process performance
Everybody is part of a process	• 100% of every employee's time is associated with one or more ETE processes
Sense of urgency to resolve issues	• Realtime issue resolution
Less risk averse	• Understand 90% of the issues from the tip of the ice burg

Figure 4.13. Design objectives for cultural transformation.

4.6 FMO Network Architecture

Tomorrow's CSPs will become Digital Service Providers, and will measure their business performance by the margins per bit of traffic. This status calls for a DSP Platform to deliver the intelligent bits to the end customers.

CSPs must develop and deploy, such DSP platforms through 4 distinct architectures: (1) Conceptual Architecture, (2) Functional Architecture, (3) Connectivity architecture, and (4) Physical Architecture (not covered in this book as it is technology and time-based).

A unified wireless and wireline network (Tikhonov, 2013) require the best reliability and resiliency. Our design objective of 99.999% availability is aimed at building an ETE DSP Platform where failures are prevented or detected and corrected before they affect customer service. This is a revolutionary concept when pushed to the extreme, balancing CAPEX and the performance. This revolutionary concept is built upon the Co0 and Co1 as the architecture must deliver Zero-Demand Based Maintenance for restoration of all services, as well as providing a unified network infrastructure across wireless/wireline/video networks, and establishing a unified service and content development platform.

Figure 4.14 presents the scorecard for the digital services:

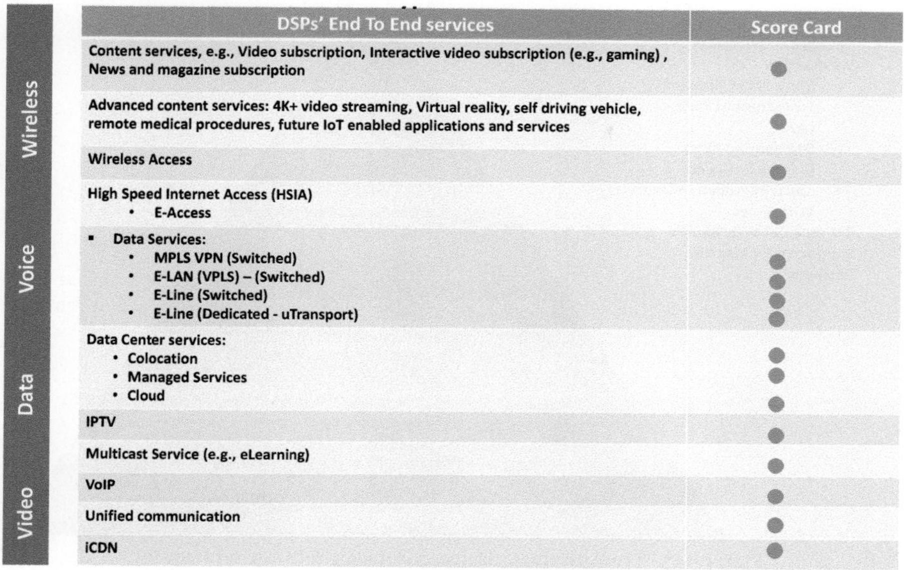

Wireless / Voice / Data / Video

DSPs' End To End services	Score Card
Content services, e.g., Video subscription, Interactive video subscription (e.g., gaming), News and magazine subscription	●
Advanced content services: 4K+ video streaming, Virtual reality, self driving vehicle, remote medical procedures, future IoT enabled applications and services	●
Wireless Access	●
High Speed Internet Access (HSIA) • E-Access	●
▪ Data Services: • MPLS VPN (Switched) • E-LAN (VPLS) – (Switched) • E-Line (Switched) • E-Line (Dedicated - uTransport)	● ● ● ●
Data Center services: • Colocation • Managed Services • Cloud	● ● ●
IPTV	●
Multicast Service (e.g., eLearning)	●
VoIP	●
Unified communication	●
iCDN	●

Figure 4.14. FMO typical services.

A world-class network survivability protocol is critical to ensure 99.999% reliability. The world-class competitive CSPs must provide a laser-sharp focus on what it takes to deliver a high availability network, with protection at each layer, engineering rules enforcing resiliency, capacity added in line with the engineering rules, and the extreme case of failures being supported by Disaster Recovery.

4.6.1 BII Survivable Network Architecture

A DSP platform must deliver a five 9's of reliability and service availability. We strongly recommend that the network architecture be designed with redundancies in each of the 6 layers of the architecture (Figure 4.15) in order to ensure the five 9's of reliability: (a) transport, (b) access, (c) aggregation, (d) routing, (e) SoIP (Service over IP), and (f) disaster recovery.

The DSPs must design and build their network under failures to avoid service interruptions.

What differentiates a BII DSP from any other CSPs is to architect the network based on the Co0 for restoration. In that,the network is self-healing and the failures present Zero "on-demand maintenance" for service restoration. The Co0, as applied to the network architecture, presents an innovative approach to how networks must be designed and built to withstand failures. Co0 ensures a near-zero impact on QoE, it lowers OPEX significantly, and it increases the CAPEX slightly to achieve the results.

This means that any major network incidents, which are counted in the 100's weekly, will have minimal to no impact on service availability for the end customers. It also means that the network resources do not need to be marshaled and expedited

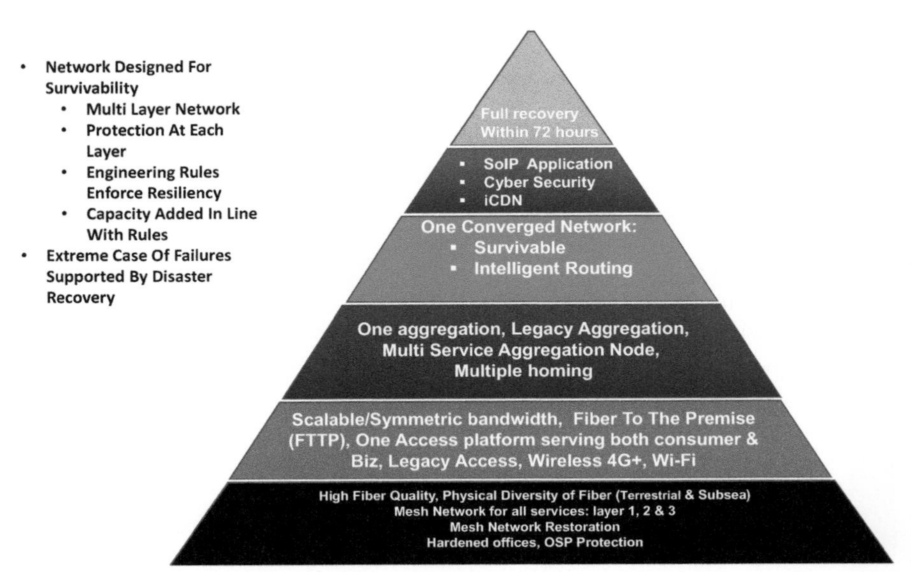

Figure 4.15. Transport access aggregation routing soIP disaster recovery.

with each incident (which otherwise has a significant uptick in OPEX) in order to resume customer service.

FMO designed for survivability with Co0 & Co1:

- Transport: One of the key and fundamental architectural issues is the physical diversity of the fiber network, in terms of the mesh network supporting the layer 3 IP network, as well as point-to-point connectivity for private line services. It is key to ensure that fiber connectivity among the network nodes are diverse and NOT collapsed (meaning that the recovery path is not riding on the same fiber sheath as the original path).

  At the base of the transport network,we must ensure the physical diversity of the fiber network,both undersea as well as terrestrially. If DSP owns this asset, it must proactively plan this—where DSP is leasing fibers, must aggressively check and verify for physical diversity.

  Terrestrial fiber is preferred underground vs. overhead. It is critical that the central office locations are designed to withstand hurricanes, floods, etc. The best test is that if the COs are fortified in such a way that, for example, National Emergency Organizations (e.g., FEMA)would prefer to set up its HQ in these locations in case of a disaster.

  Using these physically diverse connections, DSPs can interconnect their transport assets to provide a resilient and a managed bandwidth capability for customers and the higher-level service platforms. DSPs must be building mesh networks globally to provide maximum resiliency and must undertake significant certification and testing of all equipment to assure it operates with appropriate reliability.

Also, it is critical to provide proactive outside-plant-protection (OSP) of the transport assets. The OSP operations must provide asset surveillance, and "one-call" support to protect the transport network.

The protection in the transport must be provided through mesh connectivity, with no collapsed fiber.

- Access: the access network must provide symmetric/scalable bandwidth and, where needed, premise fiber and power diversity, redundant access through wireless/WiFi access, and dual and diverse homing to the network. It is a significant CAPEX concern for access to be supporting both consumers and businesses on the same unified access platform.
- Aggregation: IP/Ethernet aggregation (minimum a two-level aggregation) with dual chassis and dual-homing is key for network survivability.
- Routing: Failure resistant, dual chassis, Intelligent Routing is the driver for survivability in the layer 3 network.

  At these service layers, DSP must design and build the network with multiple homing. This means that the traffic from one location has the option of being passed to at least two other locations for onward transit—and by building the connections on the robust Layer 0/1 network, there are significant resiliency options in place that minimize the need to use the multi-homing capability; however, if there is a failure on a link, then the routing equipment will exploit intelligent routing algorithms to move traffic through an alternate link and maintain customer service. DSP must assume a worst-case scenario and build sufficient capacity in and out of the network to carry all traffic down alternate paths—this 'rainy day' approach ensures that the DSP's customers experience sunny day service in the event of failures.
- SoIP: Service over IP (SoIP) is exploding. This direction is influenced by the lifestyle change such as blogs, social networking services, interactive video such as gaming, music and video streaming, VR, AR, robotics, etc. As a result, it is forcing the service development and networking technology trend toward convergence for the benefit of time-to-market and CAPEX/OPEX efficiency.

  DSPs must pioneer this global convergence to reap the benefit from efficiency, and lower cost infrastructure networks. The overall architecture is inclusive of the Access Layer, Network Core layer, Media Resource layer, Service Execution Run-Time layer, Service Creation and Delivery Layer, and OSSs/BSSs layer.
- Disaster recovery: Must be Designed for catastrophic events within maximum 72 hours.

  Given that we have designed resiliency into the network, there are still unpredictable disasters that can happen—when they do, DSP must have the capability to tackle these by triggering a BII Service Disaster Recovery capability—through a BII fleet of recovery trailers and a well-practiced response team to react quickly anywhere across the globe to restore services with in maximum 72 hours.

Disaster recovery must be enabled by standard node complexes with which to build the global network. These node complexes are comprised of data nodes, service and content nodes, transport nodes, power and cooling nodes, etc.

It is also worth noting that the CSP's OSSs infrastructure must also be developed to be resilient in the case of an outage with multiple redundant IT "Class A" data centers housing the strategic IT applications, with service recovery centers defined and on standby.

4.6.2 DSP FMO Conceptual Architecture

The FMO top-level, technology agnostic, conceptual architecture for DSPs is presented in the following diagram (Figure 4.16). We established the following schematic as a reference architecture for DSPs to address the conceptual building blocks for the design of the FMO.

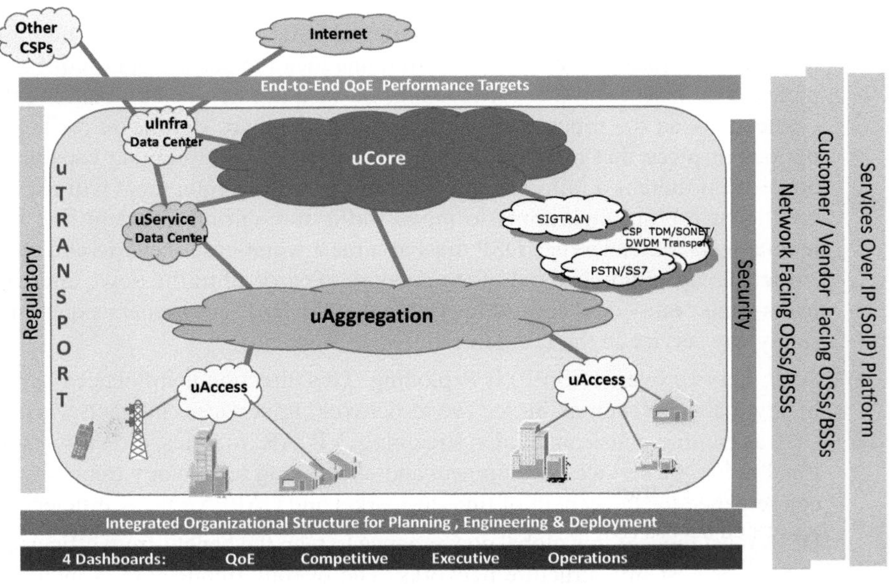

Figure 4.16. DSP conceptual FMO architecture.

This architecture is based on the industry network domains and its interconnection with the external cooperating networks.

The following key issues are addressed with this architecture:

I. **Low margin service portfolio:** expect *up to 25% improvement in the EBITDA margins through the time-to-market introduction of new services through the SoIP platform.*

II. **Not competitive QoE Performance:** expect increased *market share, and BII churn rate through migration to a unified ETE network and Operations Discipline*

III. **High CAPEX unit cost:**

   a. **Operating multiple networks:** expect *up to 40% CAPEX unit cost reduction by consolidating into One-Network core, and* a cap and grow approach on the rest of the network toward an ETE One Unified Network.

   b. **Operating unscalable legacy network technologies:** *additional and up to 70% CAPEX unit cost reduction opportunity with an all IP/Ethernet scalable aggregation and access infrastructure over the transformed One-Network.*

IV. **Not easy to do business with:** reduce *churn rate up to 50%, to BII performance of <<1%...estimated equivalent to 20% increase of revenue through reduction in OPEX.*

V. **Inefficient OPEX cost structure...***up to 80% reduction in the OPEX through extreme automation in OSSs/BSSs, zero defects, and near-zero cycle time.*

VI. **High cost of "reserve":** expect *up to 10% of the revenue in improving free cashflow and the margins.*

VII. **High cost of SLA compliance...** *up to 5% of the revenue, improvement infree cashflow and the margins.*

Below (Figure 4.17) are the key gap closures that are built into the overall architecture to transform a CSP into a DSP, leveraging the BII practices and Operations Discipline:

i. Unified Service over IP (SoIP) platform with run-time execution capability to enable creation, deployment, and bundling of IP services—a departure from redundant closed system service OSSs.

ii. Unified transport supporting all connectivity- a departure from technology/service specific transport infrastructure.

iii. Integrated access (Breuer, 2013) supporting a variety of access protocols with a common backplane and common uplink to the network- a departure from service-specific access platforms with a redundant uplink to the network, and dedicated and separate transport for each consumer and business customers.

iv. Unified IP aggregation platform (Breuer, 2013)—a departure from service-specific aggregation networks.

v. Unified core for all services—a departure from service/technology specific core networks with redundant interconnection to the cooperating networks.

vi. Unified Service Data Center (Josyula et al., 2012)—a departure from a multitude of service-specific and technology-specific data centers.

vii. Unified Infrastructure Data Center (Josyula et al., 2012)—a departure from technology-specific support-data-centers.

viii. ETE QoE performance target—a departure from islands of independent targets for each component of the network.

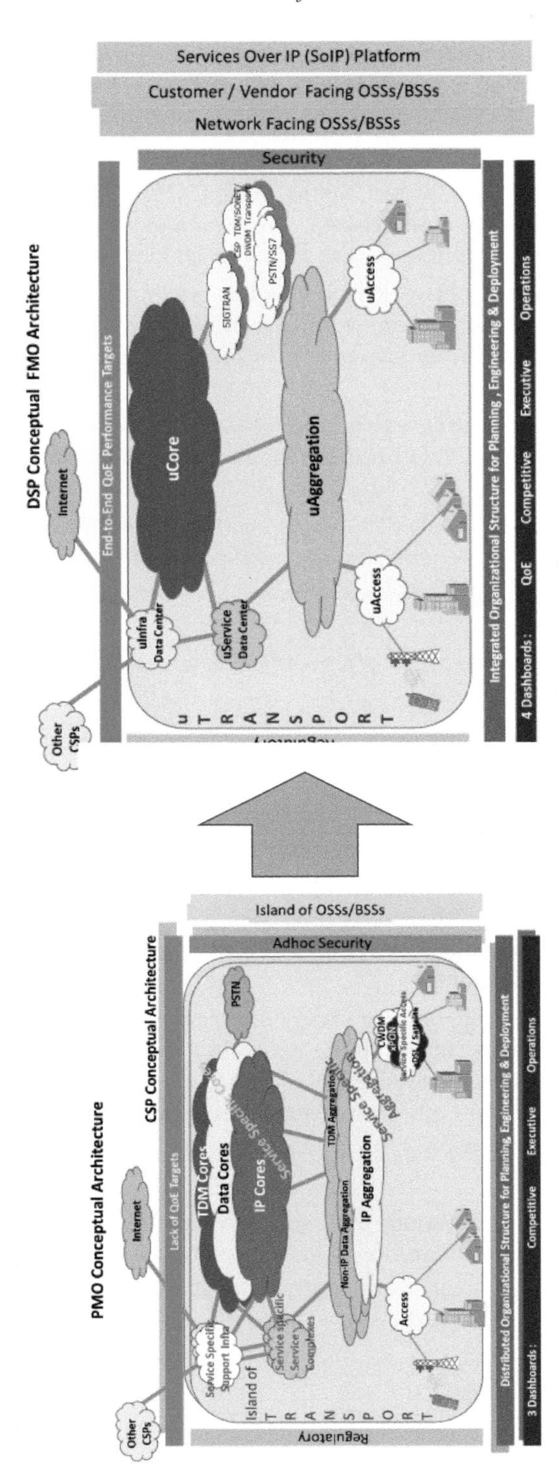

Figure 4.17. FMO conceptual architecture—gap closure.

ix. Integrated ETE organizational structure with ETE accountability—a departure from silo-based structure.

x. Integrated real-time dashboard with drill-down capability from the QoE down to root causes at network element level—a departure from disconnected non-real-time dashboards.

xi. Proactive ETE cybersecurity platform—a departure from reactive security.

xii. Customer-facing OSSs/BSSs with self-servicing capability—a departure from CSP servicing services with long cycle times.

4.6.3 DSP FMO Functional Architecture

The functional architecture provides a model that characterizes and standardizes the functions of a telecommunication system without any regard to its underlying internal structure and technology. The BII functionality of this architecture is presented below (Figure 4.18):

The functional architecture of the ETE Network will be comprised of the access network (IAN—Integrated Access Network), the aggregation network (uSAG-unified Service Aggregation), the edge, and the core network. This network will be interconnected with the Service complexes (uService) and other CSPs via the support infrastructure (uInfrastructure) complexes, etc.

Before we dive into the details of the functionality of this architecture, let's take a look at the high-level characterization of FMO's functional architecture:

uTransport is a unified Metro/Long Haul transport network supporting transport for layer 3 IP services, transport for aggregation to layer 3 IP services (IP Service Aggregation), and private line services. uTransport design must take into account a 2-plane architecture: a Dedicated DWDM plane for uCore, and an intelligent transport plane (leveraging OTN mesh) supporting access to the Layer 3 IP network, as well as private line services.

uCore is comprised of one core treated as an Autonomous System (AS) per region of the globe, and connected seamlessly, through one global control plane, to form a global network. Each AS must address the precise number of the P core nodes within the coverage of 1000 to 5000 kilometers in each region. The decision for the precise number of core nodes is critical, in that it will remain in action for decades with no change in the topology. uCore is comprised of "P" and "PE" nodes, its functionality is to perform packet processing, label processing, grooming, aggregation, and backbone routing. It is a single IP/MPLS core in one region, it has to be based on double chassis in each uCore node, it is built on dedicated DWDM unprotected transport, with TE/FRR/QoS. It must be built with adequate protection capacity, and must be designed to operate under failure. The core must be capable of sustaining a node failure, or 2 fiber span failures, to support a five 9's service reliability.

The core must be enabled with Software Defined Network (SDN). SDN enables an agile network with the creation of an open northbound open interface, this is an enabler for the DSPs to reduce TTM for service introduction, reduce CAPEX unit cost by focusing NEs to just move traffic, and reducing OPEX unit cost for network

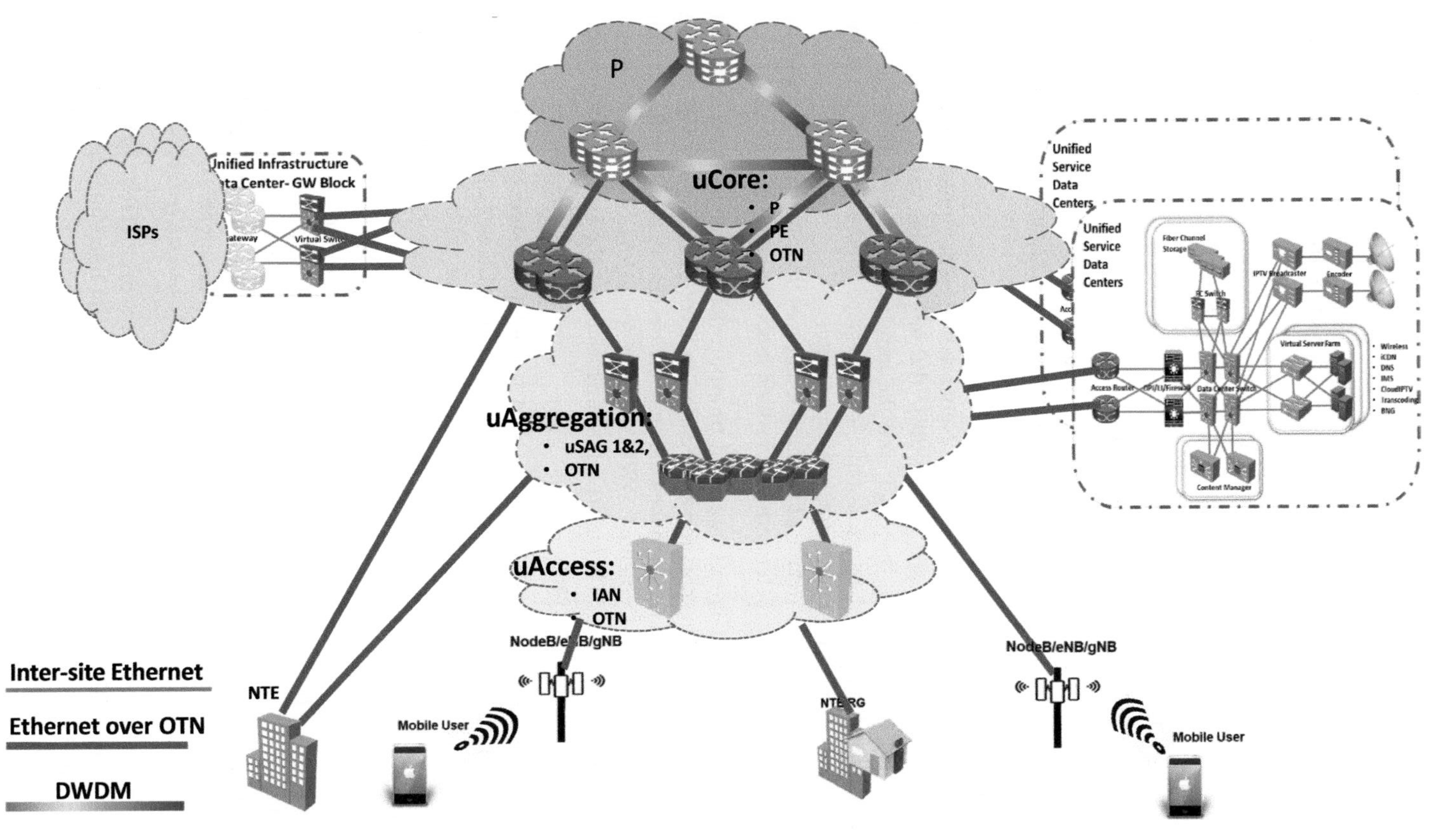

Figure 4.18. FMO high level functional architecture.

services that take a significant human capital cost to deliver, such as establishing protection, restoration or new connectivity services.

uAggregation function is to perform packet processing, aggregation, grooming, and label processing. uAggregation network is best designed for a two-level aggregation for the IP traffic, with all Ethernet aggregation, and dual chassis node complexes, MPLS enabled, and OTN mesh transport to ensure redundancy in the interconnection. Also, aggregation transport networks must be built for private line services, as well as access to the IP network. This network must leverage a partial mesh transport network with the OTN technology. Aggregation network must be able to sustain one node failure or multiple fiber span failures in order to support a five 9's service reliability.

The Aggregation network must be enabled with Software Defined Network (SDN). SDN enables an agile network with the creation of an open northbound open interface, this is an enabler for the DSPs to reduce TTM for service introduction, reduce CAPEX unit cost by focusing NEs to just move traffic, and reducing OPEX unit cost for network services that take a significant human capital cost to deliver, such as establishing protection, restoration or new connectivity services.

uAccess function is to provide protocol conversion from customer-facing to network-facing, aggregation, backhaul to uAggregation. uAccess network must be designed for Multi-Service Access, including Ethernet, xDSL + G. Fast, TDM, EPON, WiFi, 3G/4G/5G, and cable. The access backplane must be capable of being shared among all services, and the uplink must be Ethernet-based. For enterprise/ medium customers, the access must be capable of sustaining one fiber span failure, as well as an access node failure to support a five 9's service reliability. For small business customers, the access failure must be supported for repair based on SLA. For consumer customers, the access failure must be repaired given the time frame for the BII performance.

The Access must be enabled with Software Defined Network (SDN).

Unified Data Centers: The unified data centers cover support infrastructure and services. It is comprised of Unified Infrastructure Data Centers (UIDC-providing specific transport/routing functionality), and Unified Service Data Centers (USDC-providing application-specific functionality).

The data centers must be enabled with Software Defined Network (SDN). SDN enables an agile network with the creation of an open northbound open interface, this is an enabler for the DSPs to reduce TTM for service introduction, reduce CAPEX unit cost by focusing NEs to just move traffic, and reduce OPEX unit cost for network services that take a significant human capital cost to deliver, such as establishing protection, restoration or new connectivity services.

Regulatory must focus on four key elements of open market factors for success: (1) Protect customers, e.g., through public QoE dashboard and enforcement of floor level bandwidth; (2) Establish strict SLA Management, including SLA mandates and performance monitoring mandates; (3) Contain abuse of Market Power, e.g., IXP connectivity mandate to ensure QoE; and (4) Foster competition, e.g., through CSPs performance sharing mandates.

Security must focus on many aspects of cybersecurity including, (1) Network Infrastructure security, through IRSCP for uCore route black-holing; (2) Content/Service security, including LI/DPI/Filtering, distributed cybersecurity intelligence; (3) user security through CA/PKI and secure services such as VPN/SSL; and (4) SoC through CERT.

Real-time dashboards: must ensure performance monitoring and the performance targets for 5 key entities in the ETE network: (1) Access to users, (2) Aggregation, (3) Core, (4) Cybersecurity, and (5) Access to content. Also, it is critical to establish a benchmark for competitive networks.

Here the emphasis is on an integrated real-time dashboard with full drill-down capability to connect the dots from QoE to specific NEs. These dashboards include, QoE, operator's competitive dashboard, operator's executive dashboard, and operator's operations management dashboard.

Organization Structure:: The transformed network must be supported by a transformational organization structure which is focused on ETE accountability, and a new transitional oversight organizational structure to achieve the FMO.

To build a network for high availability, we must explore the uniqueness of each of the key components of this network. In the following section, we will elaborate on the construct of each component of the FMO in more detail.

4.6.3.1 Unified Transport (uTransport)

uTransport is a unified Access/Metro/LH transport network supporting transport for layer 3 IP core services, transport for aggregation to layer 3 IP core services (IPAggregation), as well as private line services. The key services for this technology are outlined below (Figure 4.19):

The transport technologies, on the Long Haul, have delivered backbone unit cost reductions of ~ 30% annually over the last 10 years. The key enablers have been the increase in the speed from OC48 to 100G+, and ULH. These trends are expected to continue through the next several years or so, and are enabled by the investment in new DWDM technologies as the spectral efficiency is promising and delivering to double and triple-digit number of channels on the same fiber with bit rates above 200 Gbps, per channel, resulting in 10's of Tbps of capacity on one pair of fiber.

uTransport End To End services	Score Card
Unprotected PE-to-PE DWDM connectivity	●
Access to Layer 3 IP	●
Private Line Unprotected	●
Private Line Protected 1+N & 1+1	●
Private Line: Dedicated Wavelength	●

Figure 4.19. uTransport ETE services.

The unit cost for the building of the transport network, in North America, has been around $100 K to $150K per mile, per fiber span. This cost is broken down into 50% construction, 25% cost of the fiber, and 25% in the cost of buildings, splicing, ROW, etc. Also, it is important to keep in mind that fiber-build average lead time is about 12–18 months.

uTransport Architecture:

The performance targets for the uTransport are outlined below (Figure 4.20):

BII uTransport Perf. Objectives	Score Card
Network Availability > 99.999%	●
95 Percentile & under failure • <12 ms latency (RTT) per 1000 kilometers customer premise-to-customer premise, to-PE • Jitter < 1 ms • Packet Loss < 0.005%	●

Figure 4.20. uTransport performance objectives.

The overall transport architecture is shown in the picture below (Figure 4.21).

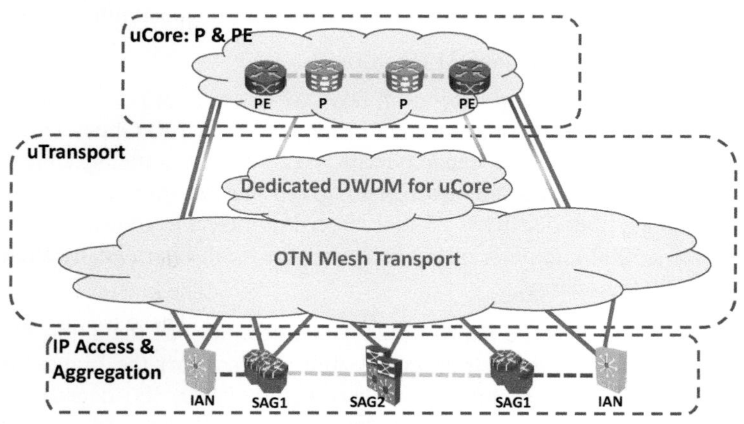

Figure 4.21. uTransport architecture.

uTransport design must take into account a 2-plane architecture: a dedicated DWDM (Sivalingam and Subramaniam, 2006) plane for uCore, and an intelligent transport plane (leveraging OTN mesh) supporting access to the Layer 3 IP network, as well as private line services.

uTransport is built on the Co1 for creation of a unified, scalable, and highly resilient layer 1 transport network for all legacy, e.g., TDM, FR, ATM, IP, and future services, including private line, backhaul for wireless and WiFi. This networking concept is to support point-and-click provisioning, grooming, and mesh restoration at the national level.

This network must be built to enable customer sites to connect to it via optical POI (Point of Interface) at customer-facing port speeds, and enable the customers to create network connections in real-time. This concept is key in providing private line services, and to enable point-and-click provisioning.

A typical uCore network with a footprint coverage in the range of 1000 to 5000 kilometers may have 10 to 40 "P" nodes and 100's of "PE" nodes that must be connected by a mesh dedicated DWDM transport. This network must be supported by 1000's of aggregation and 10,000's access nodes which must be connected by the OTN Mesh transport, and the 10,000's of business customers who are demanding Point-To-Point (P2P) private line services and/or access to a layer 3 IP network.

The mesh OTN network must be supported by 100's of OTN mesh nodes for a typical tier 1 DSP (Fujitsu, 2010).

DWDM transport plane is a partial mesh connectivity and must provide national coverage with ULH systems and ROADM to provide wavelength services for the upper layers and select enterprise customers. DWDM services must be provided for both protected and unprotected connectivity. The fiber supporting ULH should be compliant with a minimum of Enhanced SMF-28x, which is built for the lowest loss, attenuation performance, and macro-bend performance or its future incarnation. Other fiber types have transmission characteristics, such as higher polarization mode dispersion (PMD), mechanical splices and higher attenuation, which will limit their use for 100+ Gbps transmission.

Also, it is key to note that the best-in-industry practice is to place fibers in conduit vs direct buried, as new fiber is capital intensive and time-consuming to build.

(a) uTransport dedicated DWDM for uCore

DSP must deploy an intercity mesh DWDM uCore fiber network to serve key markets. The DWDM system must be capable with 100 Gbps+ (from 40 G), and double-digit channel capacity(with < 12.5 GHz channel spacing, from 80 channels spacing). For this fiber plant, given our experience, a rule of thumb for adequate fiber density is 25-to-1 (fiber-kilometers to foot-print-kilometers, e.g., for a 5000 kilometer of the footprint, the mesh fiber density should be at > 125,000 kilometers).

This fiber plant must be kept at < 15 years of aging, as the aging fiber plant exhibits fiber impairments, restricting the ability to support the highest transport speeds/ best unit cost. The major impairments include PMD, 3B1 degradation, etc.

DWDM/ROADM unprotected services must primarily be used for the Layer-3 IP network (PE-to-P-to-PE), as the IP layer protection adequately prevents failure (i.e., < 50 ms when properly configured).

It is important to note that the DWDM unprotected connectivity will be required in the aggregation network as the scope of the IP network is to be expanded to the aggregation network.

The diagram below (Figure 4.22) shows the PEs and Ps network elements connected by the dedicated, unprotected DWDM transport network. It is important to note that this is built upon the same OTN network, with the protection disabled.

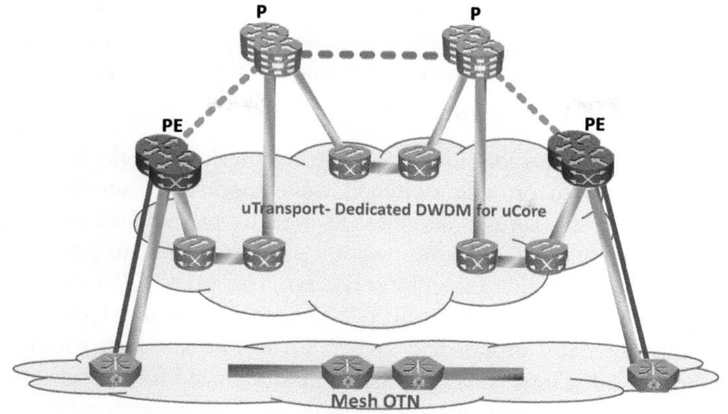

Figure 4.22. uTransport-dedicated DWDM for uCore.

Here are the architectural rules (Figure 4.23) for the dedicated DWDM transport to support uCore:

BII Architecture Rules for dedicated DWDM uTransport	Score Card
uTrasport: • Ring network to mesh network • Dedicated DWDM for uCore	●
DWDM Multi channel / > 100Gbps per channel	●
All-optical cross-connect • Multi-dimensional ROADM • CDC-ROADM	●
Zero collapsed wavelength	●
All underground	●
All placed in conduit- no direct buried • Long Haul: Owned duct	●
Partial mesh	●
No fiber with IRU	●

Figure 4.23. uTransport-architecture rules.

It is important to note that the transport network must have enough capacity under failure at peak utilization to ensure that the link utilization does not exceed the 85% threshold. Major backbone link outages run in the range of 100's per week. It is vital to not exceed the maximum capacity of the core router cards, which is under 100% (in most cases it is under 85% of the line-rate), therefore, any violation of this threshold could impact, for example, VoIP (causing > 0.4% packet loss which impacts the quality of voice, >1% causing drop calls, video clipping). Also, packet loss of > 1% will begin impacting TCP applications for page download, e.g., browsing.

The key message is to perform capacity augments to maintain a survivable network under failure during the peak traffic. It is important to deploy a modeling tool to determine unrevivable segments of the network for capacity augmentation.

(b) uTransport OTN Mesh for Private Line and Access to IP network (Figure 4.24)

The Intelligent Transport Layer is a partial mesh network. It must be built on the DWDM to provide switched wavelength/OTN services for access/aggregation to the layer 3 services, as well as private line services for the enterprise customers (i.e., point-to-point transport, point-to-point wavelength services, point-to-point Ethernet services). The OTN DWDM system must be capable of 100 Gbps+ (from 10G). The OTN switching technology allows us to electronically groom and switch lower-rate services within 10 Gbps, 40 Gbps, or 100 Gbps wavelengths, without the need for external wavelength demultiplexing and manual interconnects, and this technology enables multi-vendor interoperability for OTN switching and grooming.

These services must be provided both with and without protection. Given the scale of a typical tier 1 CSPs, the scope of the mesh footprint for these optical switches could easily run into 100's of nodes in a national geography of ~ 5000 Kilometers.

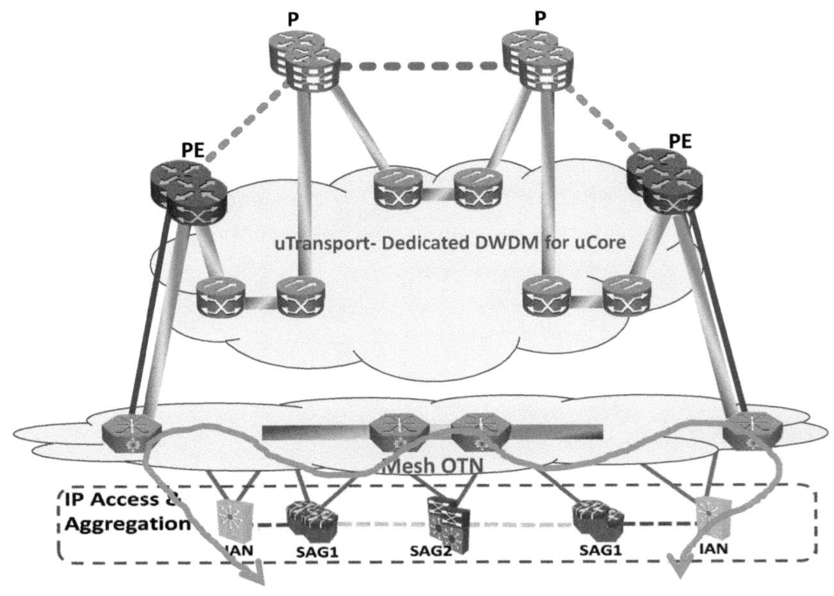

Figure 4.24. uTransport OTN mesh for private line services.

Here are the architectural rules (Figure 4.25) for uTransport OTN Mesh for Private Line and Access to IP network:

It is important to note that the regional transport unit cost has also decreased significantly, but less quickly than the backbone. The key reasons are: lower traffic volume, shorter distances between regeneration (~ 500 vs 1500 km), and the presence of poorer quality fiber on the regional routes. With the right mix of

BII Architecture Rules for OTN	Score Card
uTransport: • OTN for backhaul • OTN for Private Line	●
Zero collapsed wavelength	●
Link utilization • <80% under single failure • Engineering starts prior to 80% threshold being reached	●
All placed in conduit- no direct buried • Long Haul: Owned duct • Metro: Owned/Leased duct	●
10+ Protected Path	●
Minimal fiber with IRU	●

Figure 4.25. uTransport architecture rules for OTN.

fiber and electronics, we can upgrade the regional network to multi-channel, each at 40 G/100 G from 10 G per channel on these poor-quality fibers, if not higher.

The key question for the CSPs is how well their regional transport infrastructure is prepared for this critical upgrade? The key issues that could be impacting the upgrade are: aging fiber plant of 15+ years and fiber impairments that restrict the ability to support higher transport speeds, like PMD, 3B1 degradation, etc.

In fibers made with the 3B-1 coating (installed before 1990), the polymers break down over time, leaving a material that forms crystals. This breakdown process is accelerated by heat and humidity. As the cable cools and contracts during the Fall and Winter, the crystals cause micro-bending loss along the fiber. The loss varies linearly with temperature. Once degradation begins, it is not uncommon to see increases of 1dB per year in the seasonal loss variation, i.e., the signal strength is halved every 3 years.

There are mitigation strategies that generally involve adding equipment to increase the amount of loss that the equipment can tolerate, though there are concerns with how much dynamic range a particular technology can handle—one can design a system for large OSP loss; it's a different story to design a single system that can handle a large variation in loss from 1dB to 6dB.

Our recommendation is to characterize your fiber plant for these problems, proactively use electronic upgrade where possible, and in parallel fund multi-year programs to strategically modernize fiber in key markets for high capacity transport.

As progress is made with fiber modernization, which is key in establishing high capacity transport, the CSP will be ready to undertake "legacy ring"-to-"Intelligent Mesh" network modernization for aggregation layer transport for private line services, and backhaul to the layer 3 network by selectively upgrading or capping and growing, CSPs can evolve their networks over multiple stages to avoid any disruption to core services. The diagram below (Figure 4.26) depicts a typical legacy ring architecture.

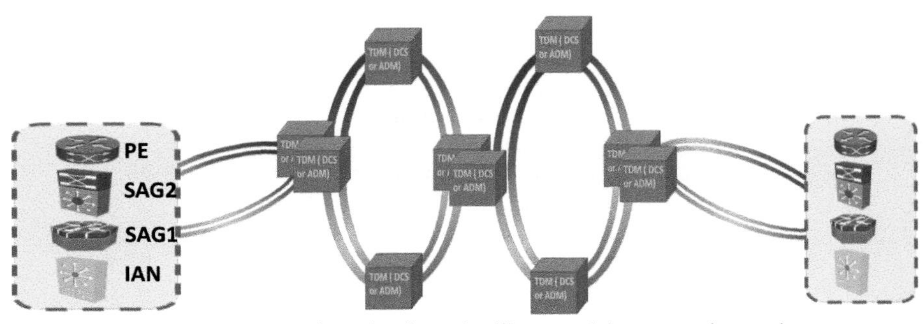

Figure 4.26. Legacy ring migration to intelligent mesh intrconnect legacy ring.

OTN provides SONET/SDH access to 100 Gb/s transport and acts as a gateway for legacy transport networks. CSPs can aim to cap the legacy transport and to grow the OTN selectively to evolve their networks over multiple TMO's to avoid any disruption to core services. The diagrams below (Figures 4.27 and 4.28) depict a step-wise approach to upgrade the legacy ring to a mesh OTN infrastructure.

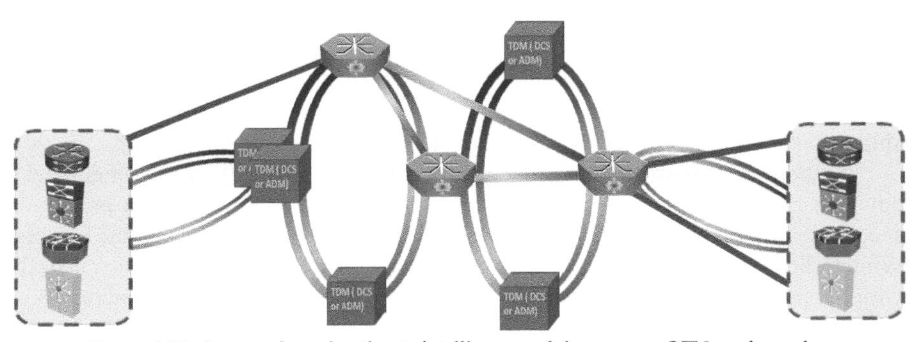

Figure 4.27. Legacy ring migration to intelligent mesh intrconnect OTN mesh overlay.

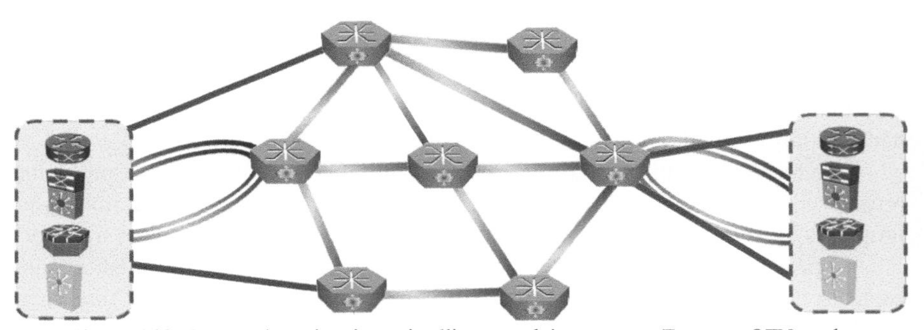

Figure 4.28. Legacy ring migration to intelligent mesh intrconnect uTransport OTN mesh.

uTransport Configuration

BII Configuration Rules	Score Card
Dedicated DWDM: OTN protection turned off	●
OTN Mesh Transport: OTN protection turned on	●
Old fiber has high attenuation	●
New fiber has bad PMD	●

Figure 4.29. Best in industry uTransport configuration rules.

uTransport Protection

BII Protection Rules	Score Card
Diversified fiber Path	●
• Negligible loss of latency	
Non-Collapsed Fiber Path • Monitored realtime	●
Minimum 3 to 10+ Protected Path	●
Last mile redundancy thru Ring or 2 P2P connections for Biz customers	●

Figure 4.30. Best in industry uTransport protection rules.

uTransport Disaster Recovery

The BII practices suggests that all UIDCs and USDCs are expected to be standardized, scalable, and built to a set specification. These data centers must be profiled, and databased throughout the DSP network. As a part of the Service Disaster Recovery process, it is expected that a series of special-purpose trailers (e.g., transport, routing, power, etc.) will be built to match the most scaled up nodes (data centers) in the DSP network, ready to be deployed to the place of a disaster. Figure 4.31 presents standard architecture for optical complexes.

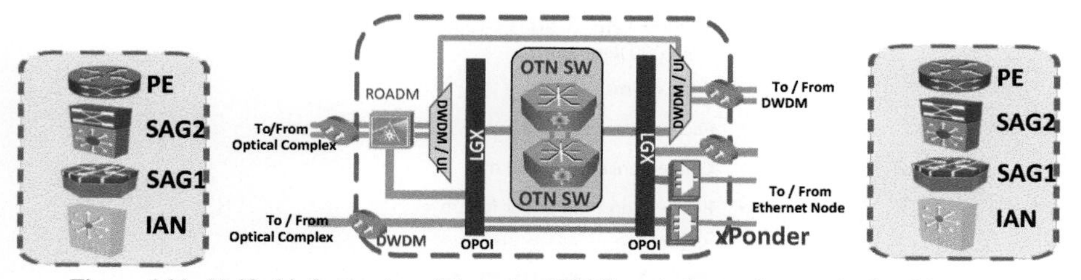

Figure 4.31. Unified infrastructure data center (UIDC) optical complex standard architecture.

4.6.3.2 Unified Core (uCore)

A unified core is a layer 3 infrastructure, shared by wireline and wireless services, and will be supporting quad-play connectivity services, dominated by VOD, broadcasting video and a variety of similar services. The uCore is a national core, or in some cases may also be comprised of a metro core connected to the national core.

The BII practice of establishing a metro core, is fueled by the need (efficiency and QoE) to keep the metro traffic local, aggregate the traffic, and leverage the national core to transport inter-region.

uCore scope is PE-to-PE, and over time, the MPLS boundary must extend, beyond the PE, to the Aggregation layer. uCore node functionality is best to focus on (1) Traffic Aggregation, (2) Grooming, (3) Label Processing, (4) backbone routing, and for the Edge node, additional functionality will include (5) SoIP (e.g., IMS)/ iCDN/IXP. The overall uCore architecture is depicted in Figure 4.32.

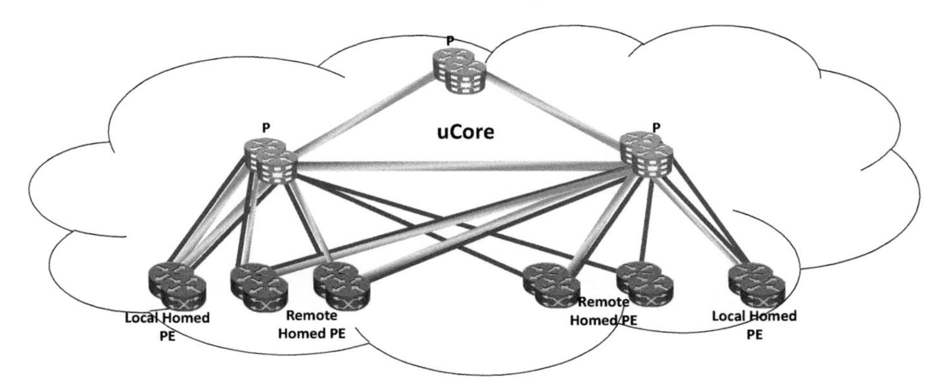

Figure 4.32. Unified national core.

The performance objective for the core network is to deliver on the following ETE performance under failure and during peak traffic (Figure 4.33):

BII uCore Performance Objective	Score Card
Network availability > 99.999%	●
Zero-Demand Based maintenance	●
95 Percentile & under failure	●
• Geographical coverage; 1000 to 5000 Kilometers (PE-to-PE)	
• PE-to-PE Latency (RTT) < 12 ms per 1000 Kilometer	
• Jitter < 1 ms	
• Packet Loss < 0.005%	
• Convergence time < 50 ms	
• Restoration time under failure < 50 ms	

Figure 4.33. uCore performance objectives.

The high-level transformational guideline for the uCore IP network must address (1) Architecture, (2) Configuration, (3) Protection, and (4) Disaster Recovery.

uCore Architecture:

Figure 4.34 presents the BII overall architecture and the associated design rules for uCore.

BII Architecture Rules	Score Card
IP/MPLS Core, extendable to Aggregation	●
PE-to-PE Latency < 12 ms per 1000 Kilometer	●
Number of P nodes decided under failure:	●
• Lowest worst case latency	
• Minimum variation in latency	
• Minimum Capex for backhaul P-toPE	
Network isolation < 3% of traffic	●
Physical location of P nodes:	●
DWDM for IP Core Nodes	●
No Optical Protection for IP core	●

Figure 4.34. uCore architecture rules.

To deliver on the performance objectives as outlined above, the converged network (Cisco, 2016) must be capable of (a) MPLS switching traffic on the common backbone, (b) the core must be internet route free, i.e., all traffic through the core must be label switched with no exception. This is critical for the stability of the core as there will be no native IP switching of customer data in the core, (c) regional core fully connected to the national core to improve QoE and to reduce CAPEX, and (d) the control plane must be able to protect against network meltdown, regional failures, and loss of major metro cities in the network, (e) must implement Class of Service (CoS) at the edge, and Quality of Service (QoS) through the core backbone to deliver consistent, differentiated service across the global converged network, and (f) implement MPLS Traffic Engineering (TE) and MPLS Fast Reroute (FRR) to provide fast, Sonnet like restoration for link and node failures in the core. The uCore must support multicast.

A converged IP network for data, video, and voice, must allow the customers (specifically enterprise customers) to establish CoS at the service edge to enable them to configure ETE QoE for their services. It is advisable to introduce at least 4 CoSs, to handle video and voice, high priority data, and best-effort traffic.

Figure 4.35 presents the rules for QoS.

Quality of Service (QoS) Rules	Score Card
MPLS Quality of Service: • Class of service at edge • ~4 Classes of service	●
Bandwidth guarantee for premium services	●
Priority treatment of control management messages	●

Figure 4.35. uCore quality of service rules.

It is critical to implement a pollution control mechanism at the edge in order to manage misuse of core bandwidth by either inadvertent or malicious setting of the CoS setting at the ingress to the edge network. The pollution control mechanism must include validation of the CoS per the customer contract.

When there is no congestion on the backbone trunks, there are no requirements for traffic management. However, during congestion, traffic with the lowest CoS is likely to be discarded. Any congestion on the customer egress links must be handled based on customer CoS markings which will be preserved across the core backbone.

Another key implementation mechanism is the deployment of Fast Reroute (FRR) for the hitless restoration of QoE in case of link and node failures. This restoration is local, in that in case of a link failure, an immediate upstream node will detect the failure and switch the traffic to a pre-configured path. This restoration speed is fast and under 50 ms, which is not service impacting.

Also, it is important to deploy traffic steering (TE) to send traffic around network congestions until capacity is deployed at the right place.

It is also of the utmost importance to implement an Intelligent Routing and Service Control Platform (IRSCP) to replace legacy Route-Reflector (RR) with significantly improved route management functionality, e.g., enabling real-time black-holing of bad traffic such as DDOS attacks through IDS (intrusion detection system) for determination of malicious traffic (attacks such as buffer overflow, viruses, worms, DOS, network scanning, etc.) using "traffic signatures" to detect the attack, and to leverage the data guards built into the IRSCP to filter out the bad traffic.

The IP/MPLS core must be equipped with control plane resiliency in order to protect against network meltdown, regional failures, and loss of major metro cities in the network. The traffic trend in the core is doubling every other year for the foreseeable future. This traffic growth is driven by the demand from the access, as well as the via (transit) traffic in the core.

A stable and unit cost-efficient core must be supported by intelligent content distribution, multicast for content distribution, and smart peering arrangement.

The Core must maintain restoration capacity for maintenance and failure scenarios. The restoration capacity is not in use most of the time and is engineered to handle primary traffic plus the largest failure traffic.

The restoration bandwidth is also a significant resource for incremental revenue generation. It must be leveraged as a service differentiation solution for paying customers. For example, CDN service or new Best Effort services (such as software download) could utilize restoration capacity and allow for better service "fill" of network capacity during the sunny-day operations. Additionally, during the rainy-day operations, the restoration capacity must be prioritized to service primary traffic.

In some unique situations, the augmentation of the core with metro/regional layer 3 networking may prove to be essential, only when it improves the performance latency. Our BII practices are indeed to allow for MPLS core per city/region, connected back to the national core. This way, with metro level core, intercity transport will create regional express links avoiding the national core. In such cases, the BII practice is to allow for > 2 points of connectivity between the metro and the national core.

In a converged network, the core cannot fail. The core must be equipped with the capabilities for outage/maintenance hitless or sub-second restoration, i.e., < 50 ms.

The off-net traffic is expected to grow significantly relative to the on-net traffic. The ratio of 2 X to 1 X every other year is proving to be normal…this projection will be critical for proper placement of the iCDN, as well as the peering points.

It is key to implement Multicast in the core to enable efficiency in the video broadcasting content. uCore is a layer 3 network, and is comprised of PE and P routing complexes. This network could easily get congested with broadcasting video, if the multicast is not implemented in the core.

uCore (PE-to-PE) must be designed for a latency of < 12 ms per 1000 kilometers, and near-zero variation in latency under failure. uCore must support hitless maintenance and < 50 ms restoration and convergence time. This calls for rapid convergence (< 50 ms through FRR (Cisco, 2020) capability), and multicast. TE (Cisco, 2009) must be extended to PE, and the aggregation routers. Also, another key BII consideration is non-stop routing with Route Processor (RP) failures, as well as hitless RP failures. As a part of the P maintenance strategy, we must also implement In-Service-Software-Upgrade, which is critical for reducing the convergence and maintenance DPM.

Given a converged uCore for quad-play connectivity services, we must build redundancy in the control plane. Ideally, complete redundancy requires two completely diverse networks to be built, with diverse vendor equipment, routing protocols, operational tools and processes, backbone links, etc., which is cost-prohibitive. The BII solution is to separate the control plane from the data plane. The cost of the core is in the data plane. Therefore, by building the redundancy in the control plane, the unit cost is well managed, while building the necessary resiliency in the data plane.

uCore must be scalable to 100+ of Tbps per node. In the absence of this scalability, the only way to expand the processing power of a core node routing function will be with a back-to-back connection of less scalable routing elements that reduce the efficiency by as much as 70%.

The number of nodes (P nodes) in the uCore is to be decided through extensive modeling, simulation, and lab testing, in that, we must analyze point-to-point latency, under 2-fiber span failures or one node failure.

The feasible number of nodes is decided when we have the lowest worst-case latency on the primary routes and up to a few secondary routes, and maximum stability (i.e., minimum variation in latency under failure). This exercise presents us with what we call a "bath-tub" curve with a feasible number of uCore nodes. Once we have established the acceptable range of number of uCore nodes, then it is essential to select the optimum number of uCore nodes where the cost of the backhaul from PE-to-P is minimized.

The physical location of the P nodes is another key factor for the transformation. The PMO "P" core locations must be seriously taken into account, the population density and future population expansion is key for the node selection. Earthquakes must also be considered as a factor in order to minimize the impact of a loss of a node. Also, another consideration for the selection of the nodes must be based on the risk associated with an attack on the national infrastructure by other countries.

uCore Configuration:

Figure 4.36 presents the uCore configuration rules:

BII Configuration Rules	Score Card
Provider Router:	●
• Dual Chassis Per Node	
• Isolation Less than 3% of Traffic	
Provider Edge:	●
• Dual Homed : Two routers, Different Nodes (If more than xx Distance)	
• Dual Homed: Two Routers, Same Node (If less than xx Distance)	
• Balanced Capacity on Homing	
Hardware Configuration	●
• High Speed Port	
• Universal Port	
Core Node Configuration	●
• PE Node, P Node	

Figure 4.36. uCore configuration rules.

uCore protection:

uCore will fail from time-to-time, however, the convergence time to recover from the failure must not exceed 50 ms as many business applications could fail/crash, and the QoE will be noticeably and negatively affected.

uCore must be designed for failures. The boundary of the failures must be defined carefully, as it is a tradeoff between the CAPEX and QoE. Generally speaking, for a five 9's of service availability, the BII protection rules under failure are summarized below (Figure 4.37):

BII Protection Rules	Score Card
Network Resiliency Failover Mode	●
• One Node	
• One Fiber Span	
• One Node – One Fiber Span	
• Two Fiber Span	
Internet route free core	●
Treat Restoration Capacity as Production Live Traffic	●
Build Intelligent Routing Service Control- e.g., mitigation of cyber attacks in the core	●
Build Network for the Peak Traffic, i.e., 95 percentile	●
Multicast capable	●
Deploy Traffic Engineering & FRR	●

Figure 4.37. uCore protection rules.

As stated above, the restoration capacity must be built into the network as production capacity, with live traffic. However, the network must be designed to allow this restoration capacity to be available to serve revenue-generating, but non-essential services. A typical service example is when a customer's application requires an OTA software update, where this service will be served as a second priority relative to the restoration application.

Given our experience having built and operated global tier 1 DSP networks with BII QoE performance, the following failures adequately protect the QoE and optimize the CAPEX investment. In that, uCore must sustain full operational performance under one node failure, or two fiber span failures. Also, the core mustn't be severely isolated under 3 fiber span failures- the maximum allowable isolation (locked traffic) must not exceed 3% under 3-span failures.

We need to ensure a balanced capacity on all redundant uplinks PE-to-P. The core is best designed partial mesh with a 1.5 load factor (average hops from P-to-P), and maximum link utilization in a sunny-day operation must not exceed 65% and must be under 85% utilization under failure as defined by the failures above.

Next, we turned our attention to node failure and the choice between 1-router vs 2-router per node architecture.

Statistics from core failures, in the tier 1 CSPs, indicate that 75% of the failures are caused by fiber cuts, and 25% of failures are caused by maintenance activities and/or Plant Operating Errors (POEs). Given the fact that the probability of POE is significant, networks must be designed to heal themselves in such cases. Probability of a complete simultaneous 2-node failure due to catastrophic events is negligible, and best is to deal with such events (i.e., more than one node failure) leveraging disaster recovery methods recommended for uCore as well as other nodes in the network.

Dual chassis routing nodes in the P and PE node complexes provide superiority for protection against network maintenance activities (e.g., software upgrade)and also enable continuous operation under POE.

Given our transformation experience and the extensive modeling, in a dual chassis configuration, when one chassis failed, the latency remained unchanged, however, in a one chassis configuration, when the single chassis failed, the latency was higher by > 20% due to inter-node complex traffic rerouting under failure.

Also, dual chassis configuration presents a potential for cost reduction for intra-office connectivity vs inter-office, or PE uplink to local P node vs remote P node. Generally speaking, local PEs are best to home on the dual chassis, and the remote PEs on two separate P node complexes.

Furthermore, to achieve the design objective for convergence time of < 50 ms, the core routing must be designed to operate under automated Traffic Engineering and steering, and fast restoration.

uCore Disaster Recovery:

The BII practices suggest that all UIDCs and USDCs are expected to be standardized, scalable, and built to a set specification. These data centers must be profiled and databased throughout the DSP network. As a part of the Service Disaster Recovery process, it is expected to build a series of special-purpose trailers (e.g., transport, routing, power, etc.) to match the most scaled up nodes (data centers) in the DSP network, ready to be deployed to the place of a disaster. Figures 4.38 through 4.40 present the standard architecture for a uCore node based on the following standard node complexes:

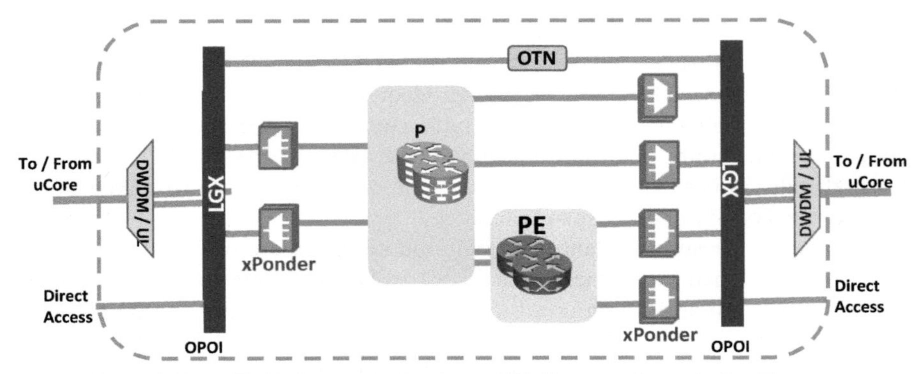

Figure 4.38. Unified infrastructure data center (UIDC) core node standard architecture.

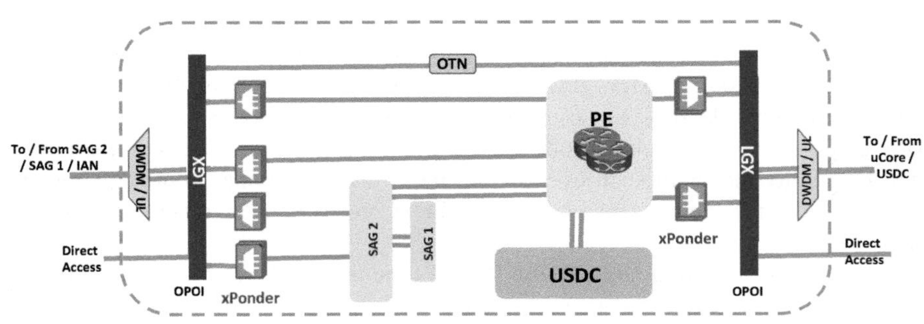

Figure 4.39. Unified infrastructure data center (UIDC) provider edge node standard architecture.

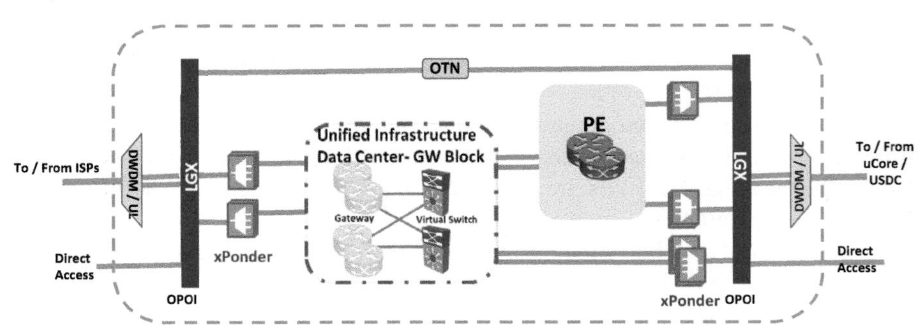

Figure 4.40. Unified infrastructure data center (UIDC) gateway node standard architecture.

4.6.3.3 Unified IP Service Aggregation (uSAG)

Aggregation network (Breuer, 2013) and (Juniper, 2015) will be supporting quad-play connectivity services, dominated by interactive video and a variety of like services. uSAG (Lange, 2010) node's functionality is best to focus on (1) Traffic Aggregation, (2) Grooming, (3) Label Processing, and (4) Packet Processing.

uSAG network is a layer 2/3 IP network, and the backhaul to the layer 3 IP network. It is comprised of IP Service Aggregation node complexes.

uSAG scope (Figure 4.41) is from the uplink from the uAccess node to the uplink to the PE nodes. It is highly recommended that the MPLS boundary be extended, beyond the PE, to the uSAG layer.

The performance objective for the uSAG (Figure 4.42) is to deliver on the following ETE performance under failure and during peak traffic:

Like uCore, the Aggregation IP network must be designed for a latency of < 12 ms per 1000 kilometers. Given the optimum coverage of ~ 300 kilometers, the round trip latency is expected at < 4 ms (this target is established based on 10 ms latency per 1000 kilometer of fiber, and .2 ms latency for every piece of network elements within the scope of the uSAG), and near-zero variation in latency under failure. Besides, the convergence time in case of failures must be < 50 ms.

The high-level transformational guideline for the aggregation IP network, must address (1) Architecture, (2) Configuration, (3) Protection, and (4) Disaster Recovery.

uSAG Architecture:

The aggregation IP network architecture must lend itself to aggregate low bit rates from the customer sites to the high bit rates for hand-off to the uCore and/or regional core. The most economical architecture must be based on a minimum of 2 level aggregation: Unified Service Aggregation uSAG1 and uSAG2. The intent is to optimize the CAPEX by enabling the lower bit rate connections to get aggregated in two levels before delivery to the higher unit cost ports on the PE nodes at 40+ Gbps.

The architecture must preserve aggregation from legacy access, such as Sonnet and TDM, while providing high-speed downlink (customer-facing) Ethernet services of 1–10 Gbps for SAG1 and 10+ Gbps for SAG2. It must provide integrated aggregation routing capability through the single backplane, 10+ Gpbs uplink ports, and CWDM/DWDM downlink ports. The key feature set must include QoS/CoS marking, EoMPLS/VPLS for PPPoE, and full feature MPLS.

Here is the overall BII architecture and the associated design rules (Figure 4.43) for uSAG to perform at five 9's of reliability.

uSAG Configuration:

uSAG must take into account scalable Ethernet universal port with dual chassis, with a balanced capacity for downlink and uplink. Dual homed from SAG1 to SAG2, and PEs. It is key that the architecture be non-blocking as this network must support "symmetric communications". uSAG configuration rules are presented in Figure 4.44.

uSAG Protection:

uSAG must be enabled with FHRP in SAG 1 and PE for business customers. Both uSAG1 and uSAG2 must be MPLS enabled, and adequate capacity must be provided in order to support constant bit rate plus bursty traffic, e.g., uSAG1 for 1:4, and uSAG2 for 1:3, with 1:2 for connection to PEs. It is also critical to plan for protection capacity on dual-homed links. uSAG protection rules are presented in Figure 4.45.

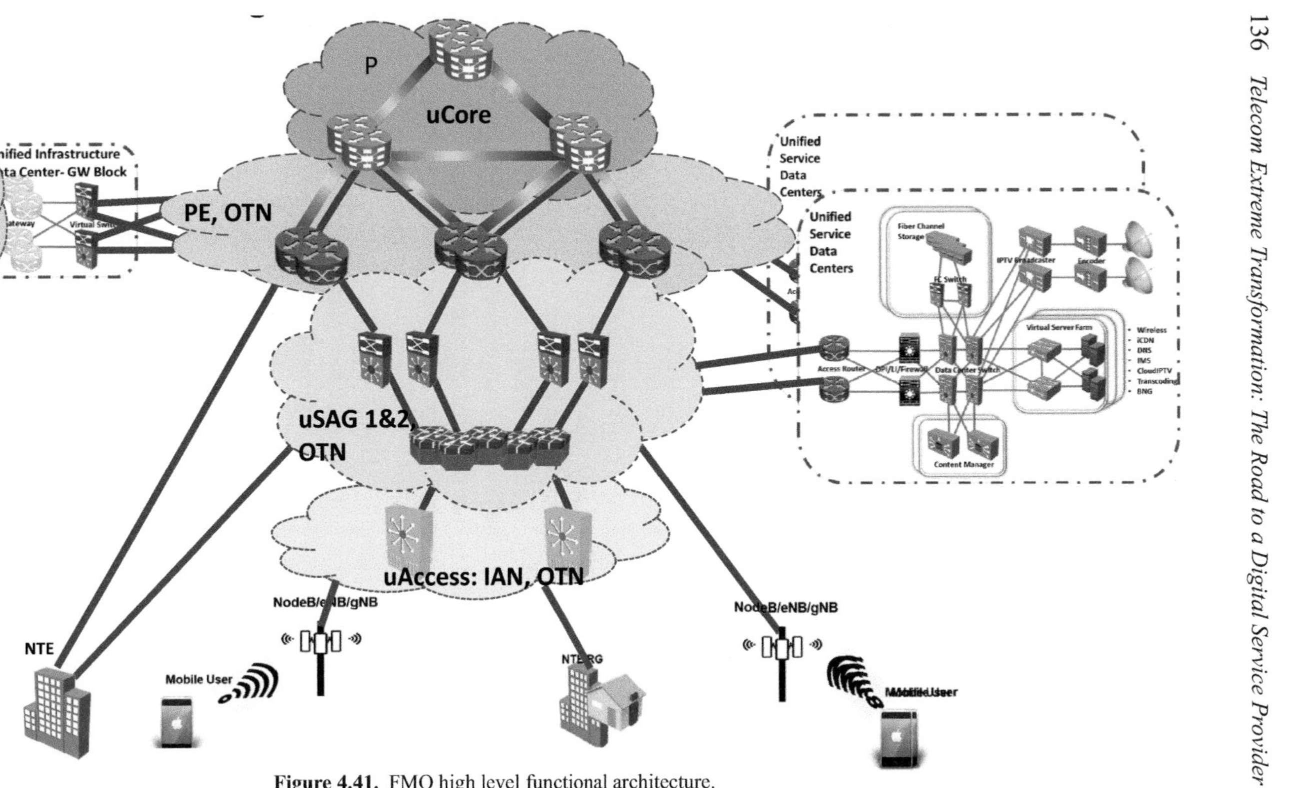

Figure 4.41. FMO high level functional architecture.

BII Aggregation Performance Objective	Score Card
Unified access network for quad-play services @ 99.999% availability	●
Zero demand-Based maintenance	●
95 Percentile & under failure	●
• Access Node-to-PE Latency (RTT) < 3.5 ms per 300 Kilometer	
• Private Line Access Node-to-Access Node Latency (RTT) < 12 ms per 1000 Kilometer	
• Jitter < 1 ms	
• Packet Loss < 0.005%	

Figure 4.42. uSAG performance objective.

BII Architecture Rules	Score Card
No single Point of Failure:	●
• Redundant Chassis for SAG1 & SAG2	
• FRR/FHRP for redundancy	
VPLS Service Aware	●
10GE+ universal customer port connection	●
MPLS Enabled: TE, FRR, VPLS	●
Preserve Legacy Uplink Aggregation	●
Non-Blocking Architecture	●
QoS / CoS marking	●
MPLS Multicast	●

Figure 4.43. uSAG architecture rules.

Best-in-Class Configuration Rules	Score Card
Chassis Configuration:	●
• Redundant Chassis for all Nodes	
• Single backplane Multiple Service (preserve Legacy Access)	
Scalability: Gigabit Ethernet Uplink ports	●
Legacy compatibility with downlink ports	●

Figure 4.44. uSAG configuration rules.

Best-in-Class Protection Rules	Score Card
Redundancy by FHRP in SAG 1 & PE nodes for business customers	●
Dual homed from Access node thru PE + Protection Capacity	●
Reduced convergence time by using V-Ch in SAG 2	●
MPLS enabled node, FRR	●
Mesh fiber with OTN	●
Link utilization < 50%	●

Figure 4.45. uSAG protection rules.

uSAG Disaster recovery

The BII practices suggest that all UIDCs and USDCs are expected to be standardized, scalable, and built to a set specification. These data centers must be profiled and databased throughout the DSP network. As a part of the Service Disaster Recovery process, it is expected to build a series of special-purpose trailers (e.g., transport, routing, power, etc.) to match the most scaled up nodes (data centers) in the DSP network, ready to be deployed to the place of a disaster.

Figures 4.46 and 4.47 present standard architecture for uSAG complexes, which inludes uSAG2 and uSAG1. The node interconnections are all supported by OTN with optional layer 2 protection or FRR.

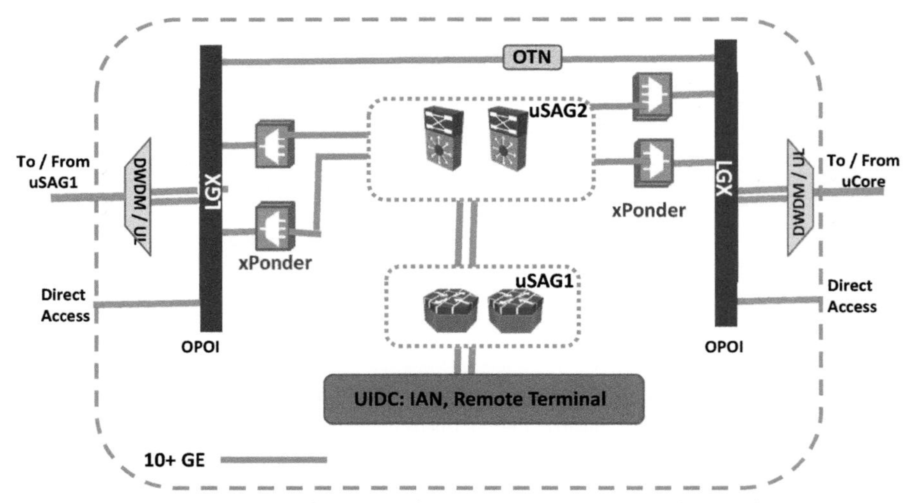

Figure 4.46. Unified infrastructure data center (UIDC) SAG2 standard architecture.

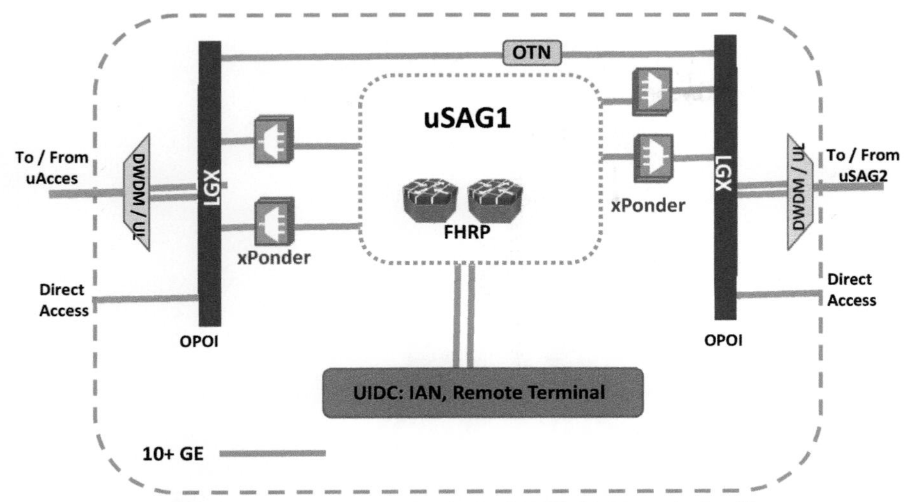

Figure 4.47. Unified infrastructure data center (UDIC) SAG1 standard architecture.

4.6.3.4 Unified Access Network (uAccess)

Unified Access network supports quad-play connectivity services, dominated by interactive video and a variety of like services over wired and wireless media. The complexity of managing separate wired data, wired video and wireless networks, multiple management systems, multiple network operating systems, and chaotic device on boarding processes is the driving force behind the unification of the access network. Access node's functionality is best to focus on (1) Aggregation function, (2) Grooming, (3) Backhaul to Aggregation IP Nodes, and (4) Backhaul to aggregation Private Line nodes.

uAccess connects wireline subscribers and wireless access points/cell sites (Adtran, 2020) to the immediate UIDC in the DSPs' network, and its scope includes Customer Premise Equipment (CPE) and SIAD for wireless, CSPs' NTE/ONU collocated at the customer location,the Integrated Access Node (IAN) in the UIDCs, and the backhaul network connecting the uAccess devices. To preserve the legacy access, it is always the case that the existing FTTN, IPDSLAM, EoCU nodes are IAN upgradable.

The IAN performs edge packetization (for PSTN, TDM grooming, and TDM-to-IP conversion function, and pass-through for IP NTE). It hosts various access technologies and can be expected to be deployed with the kind of technology that makes the most sense in the location it is placed. For example, IAN in the UIDC may have functionalities including, ADSL2+, SHDSL, and PON. IAN at the SAI may have VDSL2 and optical Ethernet. The benefit of IAN is that it can reduce the disparate chassis, power, and also the trunking from the PMO as well as allowing homogenization of management, control, OAM, and trunking across services. This provides a layer of pre-aggregation and also an interconnect among services that were not previously available.

Here is the unified Access Network Architecture (Figure 4.48) for the majority of DSPs' services, including the legacy technologies.

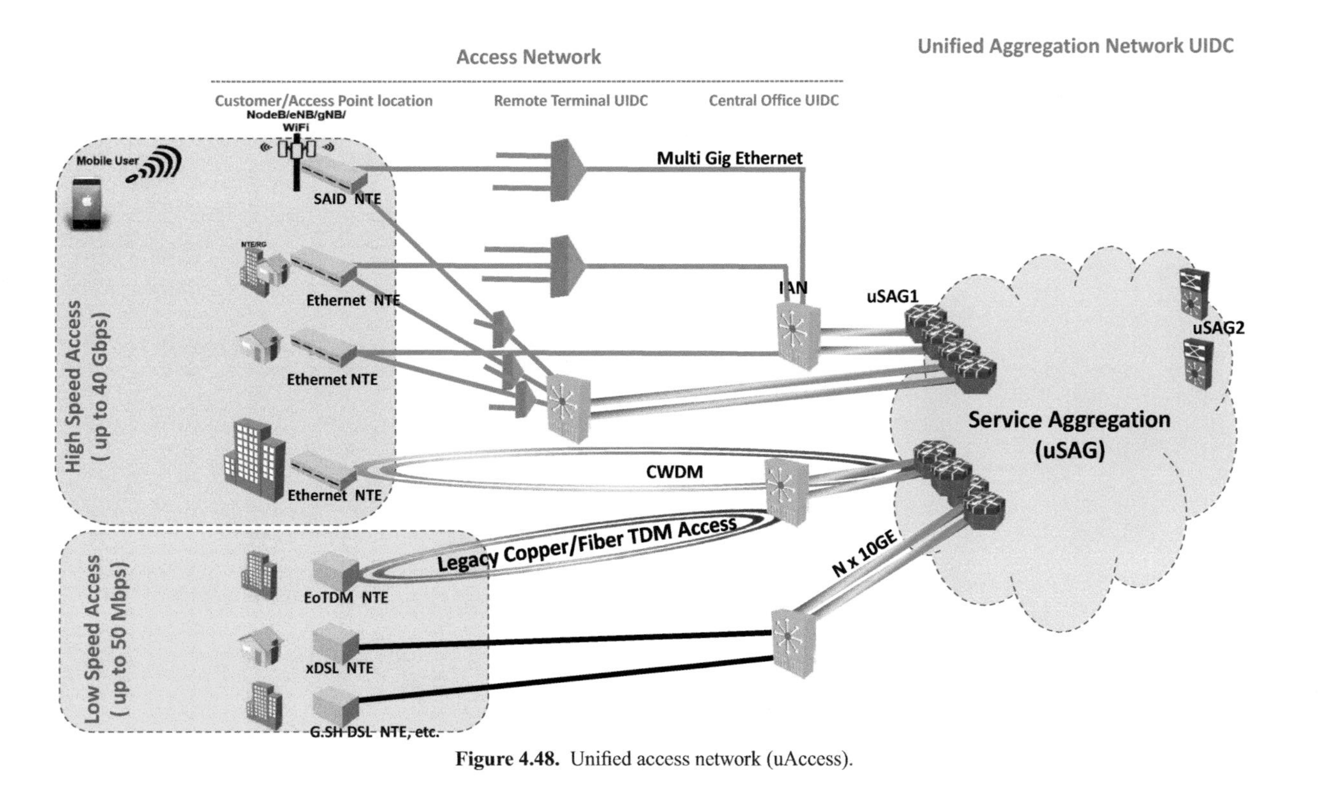

Figure 4.48. Unified access network (uAccess).

Access network is a layer 2 ethernet network, as well as a point-to-point transport for private line services, and backhaul to the aggregation network.

The access transport network must be transformed with densification to enable connectivity to the 5G+ wireless access points at high frequencies. This architecture lends itself to be configured for data tonnage reduction on the wireless access network, measured in 100's of PBs per year:

- o Push traffic to WiFi via consumer and enterprise small cell solutions.
- o Apply lower QoS for High Users to relieve congestions at cells sites during busy hours.
- o On-Device (e.g., Android) content caching program.
- o Capability in place to block the "big hitter" P2P applications.
- o Throttles data usage for prepaid subscribers that use more than the pre-defined cap thresholds in a given month.

The performance objective for the uAccess network is to deliver on the following performance under failure and during peak traffic (Figure 4.49):

uAccess network must be designed for coverage of up to 80 kilometers for xPON for Gbps bandwidth, up to 10+ kilometers for wireless, and up to 3 kilometers for xDSL for < 50 Mbps. The round-trip latency is influenced by the last-mile technologies in use, as stated above.

The high-level transformational guideline for the access network, must address (1) Architecture, (2) Configuration, (3) Protection, and (4) Disaster Recovery.

uAccess Architecture:

IAN architecture (Figures 4.50 and 4.51), when servicing xPON endpoints, must enable serving both business and consumers on the same chassis. Hence, it must be architected with five 9's of reliability, as well as symmetric bandwidth. A single platform can serve both Residential and Business/Enterprise subscribers by virtual/physical medium separation without the need for two physical solutions. The data can be carried on the same fiber, but the encryption is bi-directional and unique to every PON Port, VLAN, and LLID. This architecture is a critical capability for the DSPs to lower their fiber plant costs.

The IAN architecture is all-optical on the uplink to the network, the uplink is CWDM (8+ wavelength) with 10+ Gbps bandwidth per channel, which is essential for the scalability of the access node. The architecture must allow one backplane to be shared by service-specific, blade-based access technology to service legacy access (e.g., xDSL, EoCU, EoTDM), as well as xPON (e.g., EPON, GPON). When servicing xPON endpoints, it must have a non-blocking architecture to ensure SLA compliance (in that the switching and routing is removed from the access layer of the network, and placed in the aggregation and core networks). IAN must provide bidirectional encryption per PON Port, LLID, VLAN, meeting and exceeding the latency and jitter requirements, with zero packet loss.

IAN, in a point-to-multipoint optical network, must provide auto adjustment for the optical performance and compensation. IAN must have a software algorithm that allows for better performance on optical networks, eliminating the need to customise

Best-in-Class Access Performance Objective 95 Percentile and Under Failure	Best-in-Class Access Performance Objective: 95 Percentile & Under Failure	Score Card
EPON	• Latency RTT: 1 ms	●
	• Jitter < 0.5 ms	
	• Packet Loss < 0.001%	
SONET/SDH:	• Latency RTT: 2 ms	●
	• Jitter < 5 ms	
	• Packet Loss < 0.005%	
xDSL/EoCu:	• Latency RTT: 10-70 ms	●
	• Jitter < 5 ms	
	• Packet Loss < 0.05%	
TDM:	• Latency RTT: 2 ms	●
	• Jitter < 1 ms	
	• Packet Loss < 0.005%	
Wireless:	• Latency RRT < 20.2 ms	●
	• Jitter < 1 ms	
	• Packet Loss < 0.005%	

Figure 4.49. Unified access performance objective.

BII IAN Architecture Rules	Score Card
• Fiber multiplying technology (CWDM) on uplink at 10+ Gbps per channel	●
• Single backplane Multiple Services (Ethernet, preserve Legacy Access)	●
• Bi-direction encryption	●
• Plug & play discovery & self configuration of SLA	●
• In-line optical power compensation for extended range	●

Figure 4.50. Unified access architecture rules.

BII IAN Architecture Rules-Continued	Score Card
• One platform serving business & residential customers, Symmetric, scalable to 100's of Gbps	●
• SDN enabled for SLAs & LLID with QoS	●
• QoS: Non blocking IAN Architecture	●

Figure 4.51. Unified architecture rules—continued.

attenuation on each endpoint of the network to bring light levels from multiple sources to the same 'brightness'. By recording the incoming light into the transceiver and recording its value, next time that value is called out,the AGC (Automatic Gain Control) is pre-adjusted to the proper level instead of trying to find the proper level on its own, which is time-consuming.

uAccess Protection:

IAN is the platform for the functionality in the unified access network. It must be protected as a node (Figure 4.52), and must protect the downlink to a customer, in case of a downlink and/or ONU/NTE failure as well as uplink to the network in case of auplink and/or primary router failure in the network, this protection can easily be provided through FHRP. Therefore, it must be implemented for all IAN nodes connected directly to the network in SAG1.

Also, IAN must offer power source redundancy per line card. Each card must have two power supplies, coming from two different sources, it must have redundant EMS system which resides on each line card, in the chassis on the main controller, and in the cloud or centralized cloud servers. Each line card must come with its CPU traffic manager. This type of offering provides advanced capabilities that are required in a business deployment.

BII IAN Protection Rules	Score Card
Redundancy by FHRP in SAG 1 & PE nodes for business customers	●
Redundancy: • 2 power supply per line card • 2 Power source per line card • Redundant EMS system on line card, and in the cloud • Dedicated CPU / traffic manager per line card	●
On-demand access redundancy for consumer	●
Mesh fiber in the last mile distribution loop	●
Dual EMS: On the line card and cloud enabled EMS	●
Redundancy for end customers thru wireline and wireless connectivity	●
Last mile ring protection. Eneterprise dual fiber span feed, 1+1 for IAN line card	●

Figure 4.52. Unified access protection rules.

uAccess Disaster Recovery:

The BII practices suggest that all UIDCs and USDCs are expected to be standardized, scalable, and built to a set specification. These data centers must be profiled and databased throughout the DSP network. As a part of the Service Disaster Recovery process, it is expected to build a series of special-purpose trailers (e.g., transport, routing, power, etc.) to match the most scaled up nodes (data centers) in the DSP network, ready to be deployed to the place of a disaster. Figures 4.53 through 4.55 present standard architecture for uAccess complexes, including IAN and Remote Terminal.

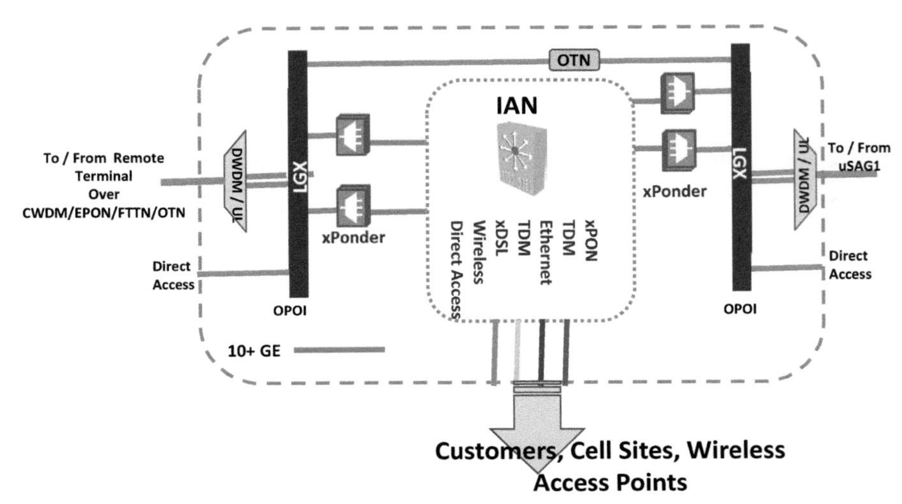

Figure 4.53. Unified infrastructure data center (UIDC) uAccess standard architecture.

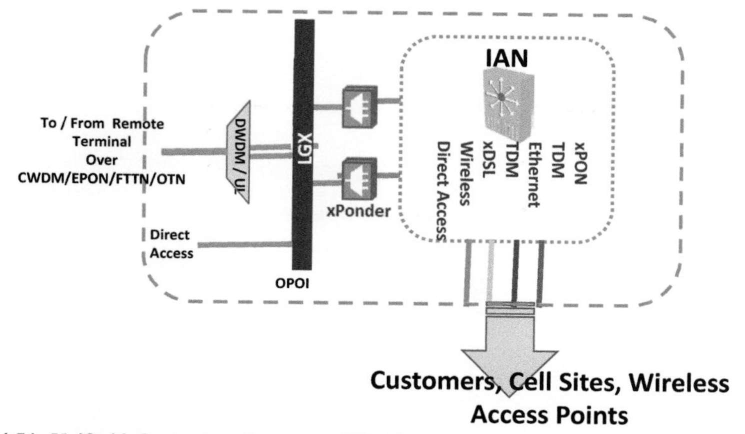

Figure 4.54. Unified infrastructure data center (UIDC) uAccess remote terminal standard architecture.

4.6.3.5 Wireless (future release Volume II)

[This section is planned to be expanded in the next release of the book. The outline below is aimed to provide an introduction to what is planned to be included in the next release.]

As we stated before, the Access traffic is growing at a rate of 30% YOY, mostly through wireless access triggered by the mobile and IoT devices. As a result, wireless access bandwidth is under significant pressure. The 5G+ wireless technologies must provide several orders-of-magnitude more bandwidth than 4G, with low single digit latency, and massive capacity consistent with fiber-based connectivity as compared with 4G.

We, the authors, aim to address this critical part of the DSP networks in much greater depth in the next release of this book. For now, we outline some highlights of the challenges involved in what is to come later: (1) unification of the wireline and wireless networks, (2) deployment of millimeter wave band, and (3) densification of the fiber network including dynamic spectrum sharing, and (4) WiFi offload.

(1) Wireless network must converge with wireline/video networks to gain economy of scale and integration. It is critical to leverage one unified architecture to gain significant CAPEX/OPEX reduction, and to enable rapid service introduction:

- Access independent IMS, and SoIP platform.
- Unified Core backbone.
- Unified Aggregation network.
- One service experience (single sign-on, single profile, different persona, single bill, and service portability across access technologies).
- Unified POPs to house wireless networking gears, e.g., RNCs, SGSN, GGSN, LTE eGW along with wirelines.
- Unified Data centers to house service gears, e.g., SoIP, DNS, Contents, Apps.
- One device equipped with multiple interfaces for wireline and wireless access networks.

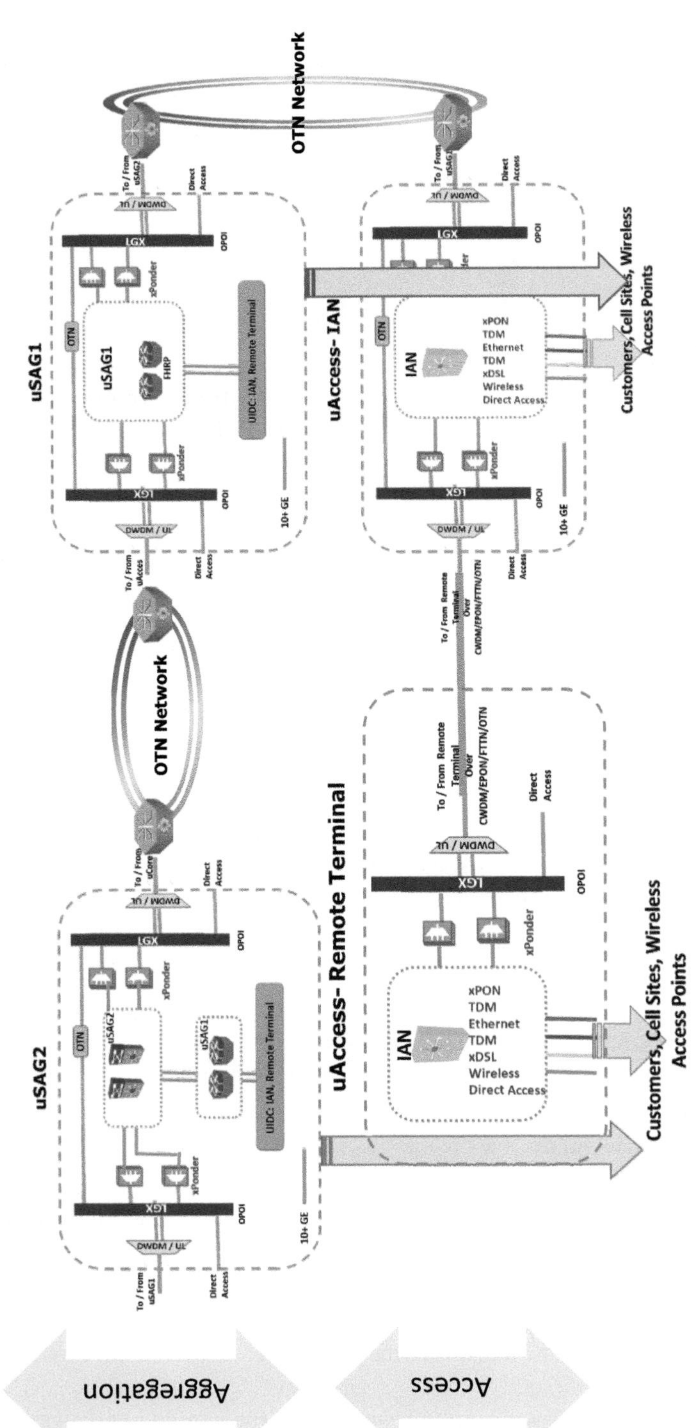

Figure 4.55. Unified infrastructure data center (UIDC) integrated aggregation & access network.

(2) Deployment of millimeter wave band to gain many orders-of-magnitude improvement in capacity and bandwidth. The mobile operators should never turn down the opportunity to acquire new spectrum. 600 MHz is great for providing bandwidth over a wide area and in hard-to-reach places. However, due to limited channel bandwidth of ~ 10–90 MHz, e.g., LTE, its capacity is too significantly limited (average of 30+ Mbps download and up to 15 Mbps upload) to handle the explosion of the video traffic.

To meet the demand of massive capacity and super-fast data rate, we have no choice, but to leverage millimeter wave spectrum (~ 24+ GHz) with multi GHz channel bandwidth vs 4G at 10–90 MHz channel bandwidth. This present significant challenge both technological and logistical.

Technological challenges are related to the "path loss" properties (path between a UE and the Base Station) and include of optical phenomena, such as aerosol induced dispersion, which takes hold and make communication at long ranges problematic, and propagation and interference impairments, including multipath propagation, doppler, shadowing, variable dielectric properties, etc.

Logistical challenges are significant, in that there is a need for densification of small cells to deploy a massive number of 5G+ access points at the national level. For example, in the United States alone, there are currently a total of 350,000 cell sites at the national level. However, a Google study for the Pentagon's Defense Innovation Board indicated that the number of the access points (transmitters) which are required to provide coverage, at ~ 100 Mbps, for roughly three quarters of the US population, using millimeter wave for 5G, is at ~ 13,000,000. This presents a massive undertaking from Capex requirements and the time-to-market perspective.

(3) Densification of the fiber network and dynamic spectrum sharing to enable transformational capacity expansion as well as high speed bandwidth. Given the challenge for the deployment of millions of 5G+ access points, there is also a significant challenge associated with the connectivity of these access points to the unified network through fiber-based transport. It is imperative for the DSPs to increase the fiber density out of their POPs, specifically in the metro areas. Also, it is as necessary to leverage the dynamic spectrum sharing technology, to use the same frequencies for 4G/5G/5G+ devices simultaneously.

(4) Deployment of seamless WiFi offload to leverage the lowest unit cost network architecture for connectivity. Unlike the wireline users with fixed location and predictable bandwidth requirements, the location of the mobile users is unpredictable. For example, in a venue complex, e.g., sporting venue, it is often the case that 10,000's of mobile broadband users come together during a short event of a few hours or days. This presents a significant cellular capacity challenge for a short period. Adding more of the cellular capacity (e.g., 3G/4G/5G+) is not economical as it requires the addition of costly cell sites/ access points, as well as acquisitions of spectrum to improve the coverage.

The choices for DSPs are limited: use macro cells with 600 MHz to < 6 GHz band for wide-area coverage and hard to reach areas, and compliment that with small cells at an unlicensed spectrum of 2.4/5 GHz with intelligent integration for making network selection and authentication to operator owned (or partner-owned) Wi-Fi networks automatic, seamless, and secure, or compliment with small cells (e.g., 5G) at the licensed spectrum of ~ 24+ GHz. Let's keep in mind that the advantage of the use of unlicensed spectrum is that the smartphones are already equipped with the chipsets to operate at those frequencies.

The above key challenges are what the authors will be covering in detail in the next release, Volume II.

4.6.3.6 WiFi Offload

As we stated, the access traffic is growing at the rate of 30% YOY, mostly through wireless access and on the mobile devices. 4G/5G wireless technologies offer higher speed and more capacity. However, as long as data usage caps remain in effect and/or the anticipated 5G higher prices, there will be an incentive for many customers to offload data access to Wi-Fi (Aruba, 2020). As a result, Wi-Fi offload is going to be higher on the 5G networks for many years to come.

One of the key issues with the wireless technology is the fact that the location of the mobile users is unpredictable. For example, in a venue complex, e.g., sporting venue, it is often the case that 1000's of mobile broadband users come together during an event. This presents a significant cellular capacity challenge for a short period. Adding more of the 3G/4G/5G capacity is not economical. However, Wi-Fi offload provides an invaluable and economically feasible complement to cellular networks in the delivery of high-quality broadband services to smartphone users.

The choices for DSPs are limited: use macro cells with 600 MHz to < 3 GHz band for wide-area coverage and hard to reach areas, and compliment that with small cells at an unlicensed spectrum of 2.4/5 GHz with intelligent integration for making network selection and authentication to operator owned (or partner-owned) Wi-Fi networks automatic and secure, or compliment with small cells (5G) at the licensed spectrum of 20+ GHz. The advantage for the use of unlicensed spectrum is that the smartphones are already equipped with the chipsets to operate at those frequencies, and the CSPs' treatment of the WiFi traffic is the same as wired, i.e., with generous data caps.

The key business drivers that we have leveraged for the introduction and integration of WiFi technology into the cellular are three-fold: (1) differentiated QoS and customer retention through the use of the best available connection, (2) incremental ARPU by tapping into non-SIM devices such as tablets and laptops, and (3) lower CAPEX unit cost for delivery of wireless internet connectivity. We have gained up to 10X improvement in capital unit cost for WiFi offload vs cellular alternatives.

In the business model for the WiFi offload, we included both SIM-enabled as well as non-SIM devices. For the SIM-enabled devices we experienced a boost in revenue. We offered "best-effort" for free without consumption limit, and offered premium services for a fee, with consumption limit. We also offered WiFi for 3G/4G subscribers who did not have an active data plan. For non-SIM devices, we benefitted from a boost in revenue, as we offered temporary usage-based (e.g., hourly, daily) and application-specific streaming services.

We strongly recommend capitalizing on the WiFi chipset that is already built into the mobile devices. Also, it is smart to capitalize on the existing authentication technologies, such as EAP-SIM and non-SIM (EAP-TLS, EAP-TTLS). Also, the WiFi access point leverages 802.1X for delivery of authentication messages. We must also push for the full interworking of 3GPP and WiFi by establishing Hotspot 2.0 and ANDSF standards (Orimolad, 2016) for (i) network/service discovery and selection in advance of association (HS2.0), and policy-driven intelligent network selection and traffic steering (ANDSF) to enable tightly coupled networks (3GPP & WiFi), (ii) Secure authentication, and (iii) enhanced security with WPA2.

The BII practice is to leverage both cellular and WiFi for mobility. The best of macro cellular networks for mobility, and the best of WiFi over wireline network for bandwidth availability and efficiency, integrated through Wi-Fi-offload on unlicensed spectrum. This integration can work seamlessly to preserve the mobile lifestyle for wireless data with much more improvement of the user experience.

The scope should include: (a) Seamless and automatic authentication with EAP-SIM & EAP-AKA, (b) hard hand-over of Non-SIM and SIM-enabled traffic routed to local internet, (c) Routing SIM-enabled WiFi traffic to the mobile core using DPI/GGSN for policy enforcement, (d) HOTSPOT 2.0 Network discovery and selection (IEEE 802.11ac), (e) Intelligent selection of best networks: 3G/LTE/5G/Wi-Fi) ANDSF in networks and devices, and (g) wireless backhaul using 5GHz mesh.

We strongly recommend a three-phase approach to WiFi offload: (1) Hard offload: WiFi Offload through local breakout, (2) Managed offload: WiFi Offload through interworking with CSP's 3G/4G/5G network, and (3) Intelligent offload: Intelligent network selection and service.

Phase 1 WiFi Offload through local breakout (Figure 4.56) enables offload to WiFI without any attention given to the quality of the WiFi connection. This method includes implementation of an interaction-free EAP-SIM authentication between the user handset and CSPs' 3G/4G/5G core, which is the same credentials used to

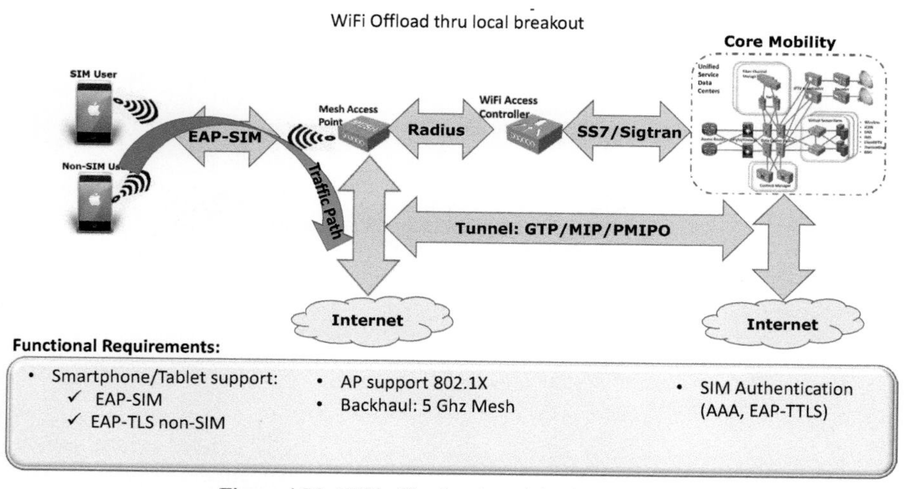

Figure 4.56. WiFi offload—phase 1 implementation.

authenticate them onto the cellular network. There is no need for a 3rd party client in the device. This approach requires a common AAA authentication server that provides 3GPP interfaces to the operator's HLR/HSS and Policy and Charging Rules Function (PCRF) to authenticate the user and enforce usage policies. The message encapsulation will be done by the APs through secure 802.1X link and RADIUS. The implementation of the access gateway and the WiFi Core Service Management Platform (WCSMP) will interwork with CSPs' core (HLR) using signaling protocol (SS7 or SIGTRAN). WCSMP will provide many functionalities, including SIM Authentication, service management (~ PCRF function), and portal support. Once SIM-authentication is complete, there will be a "hard handover to WiFi", and the subscriber will be able to use Internet service through a local WLAN (Wireless Local Area Network) breakout of internet connection. Non-SIM customers should implement authentication through EAP-TTLS servers. Non-SIM devices are mostly compliant with EAP-TLS, and this approach should offer the largest possible subscriber base.

Phase 2 WiFi Offload through interworking with CSP's 3G/4G/5G network (Figure 4.57) will enable routing of smartphone traffic for non-SIM as described in step 1, but the traffic for SIM-enabled devices will be routed to the 3G/4G/5G Core DPI & GGSN nodes for enforcement of CSPs' policies and quality of service before routing traffic to the internet.

Phase 3 WiFi offload based on intelligent network selection and service (Figure 4.58): This is the final step in the transition to a full-function WiFi offload where the WiFi is treated as another radio access technology, where this connection is intelligently selected based on the best quality link as well as service continuity between 3G/4G/5G/WiFi. This technology enables interaction between the 3G+ PCRF (policy control) and an ANDSF client on the mobile device. This network selection will be based on time-of-day, location, subscriber type, application, device type, etc.

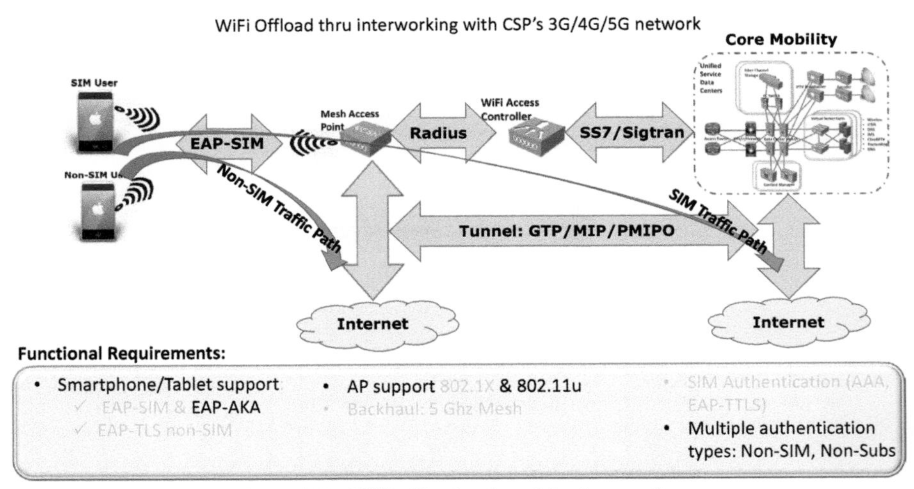

Figure 4.57. WiFi offload—phase 2 implementation.

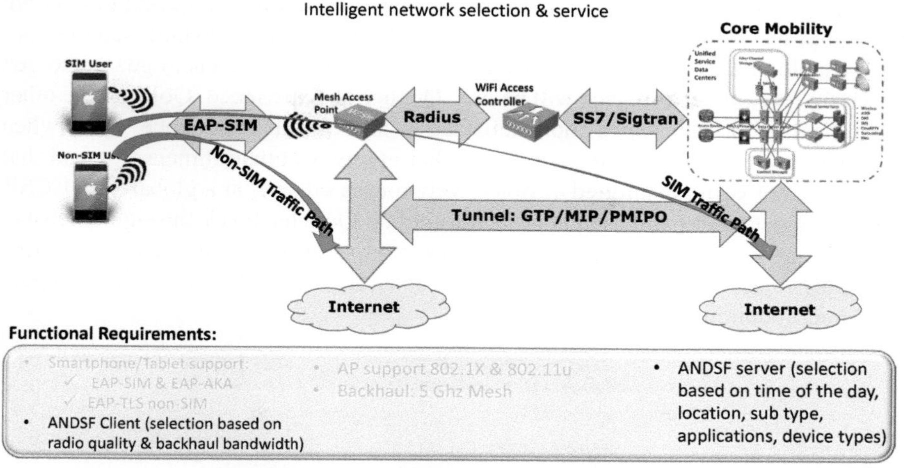

Figure 4.58. WiFi offload—phase 3 implementation.

Network selection based on the quality of the radio access and the backhaul availability is left to the device manufacturers and should be deployed upon availability.

4.6.3.7 IPTV

IPTV is a managed service for live TV content from SD resolution to 4K and up, VoD, Interactive Video, and advanced advertising for DSPs (Open IPTV Forum, 2013). Its content is controlled and protected, and its QoS is guaranteed.

IPTV shines where the video streaming option remains to be the best-effort and not reliable, otherwise, CSPs must question the viability of IPTV over the streaming alternative. The IPTV may prove to be the best option in countries where ISP networks are not reliable for service delivery (Begen, 2012).

We recommend the following QoE expectations (Figure 4.59) for IPTV:

Macro Service Flows (IPTV)	IPTV QoE Expectation	Score Card
Streaming-Service flow	• Zero pixilation • Zero freeze • Sync'd voice & video	●
Interactive Video- Service flow	• Same as streaming • Action Trigger sync'd with video	●
Bandwidth-Consumer	• Guaranteed throughput (DL/UL)	●
Service Availability	• SLA compliant • High availability	●

Figure 4.59. IPTV QoE expectation.

In the table below (Figure 4.60), we have cascaded the QoE expectation down into QoS, and will outline the BII QoS for IPTV. Quality of Service (QoS) is a measure of technological excellence as it denotes the capability of systems to guarantee that a certain level of performance will be met. Quality of Experience (QoE) on the other hand deals with user expectation, satisfaction and overall experience. QoE, when compared to QoS, is a subjective metric that involves human dimensions and that is, hence, not easily quantified or objectively measured. We, at a global Tier 1 CSP, have conducted abroad-based approach to obtain QoE feedback through extensive quantitative analysis of network QoS as we compared with QoE. Here,we outline the QoE related QoS network performance requirements that meet the stringent customer requirements:

BII Performance Objective	Score Card
Interactive Video QoE	●
• Latency < 80 ms	
• Jitter < 10 ms	
• Packet Loss < 0.01%	
Bandwidth per viewing channel: 12-18 Mbps for 4K content	●
Service Availability: > 99.999%	●

Figure 4.60. IPTV performance objectives.

IPTV must be provided over a managed network with limited geographical coverage. The IPTV services must be guaranteed with high-quality audio and video. The access mechanism is provided through a dedicated set-top box, or a better approach is through a Multimedia Application Gateway (MAG). Its content is provided and controlled by CSP. Also, the subscriber has control over the content through the parental control function.

To ensure the QoE target for availability, the resiliency of the network and the storage for the content is essential. To ensure the latency target, the network must be multicast-enabled, and protection capacity is fully deployed for the peak traffic. The packet loss target demands that FRR is implemented. For last-mile bandwidth requirements,it must be ensured that the customer profile is clearly defined for each serving market, e.g., in one market may be offering multiple 4K viewing channels per household, plus a minimum 100+ Mbps data services. This arrangement will force offering a minimum of ~ 150+ Mpbs sustained bandwidth at the last-mile, and throughout the aggregation, core, content storage, and the IXPs.

The table below (Figure 4.61) demonstrates what must be considered for the IPTV network vs the OTT streaming.

	Internet TV / OTT	IPTV
Transport	• Use best effort internet	• Use managed network
Geographical Reach	• Access available from anywhere in the world	• Access limited by the foot print of the service provider
Service Quality	• Best Effort Service	• Guarantees high quality audio and video
Access Mechanism	• Any device connected to internet	• Dedicated set-top-box
Content Generation	• No control on contents	• Managed by service provider
Content Control by Subscriber	• Not controlled	• Subscribers control contents

Figure 4.61. IPTV vs OTT streaming.

IPTV Architecture:

Figures 4.62, 4.63, and 4.64 present the architectural and the architectural rules for IPTV.

BII Architecture Rules	Score Card
Super Hub Office (Acquire national Channels)	●
Video hub Office (local channels, authentication of STB & RGW, IP Address): One office per DMA	●
Video Serving Office @ 100's (Video distribution via IAN)	●
Residential Gateway thru Multimedia Application Gateway • Open system, standard based application execution platform • Network attached Storage • DLNA compliant • Non-DLNA+ Protocols • Interfaces: USB, xDSL, Ethernet, VoIP, wireless	
Common network infrastructure: • uCore • uAggregation • uAccess	●

Figure 4.62. IPTV architecture rules.

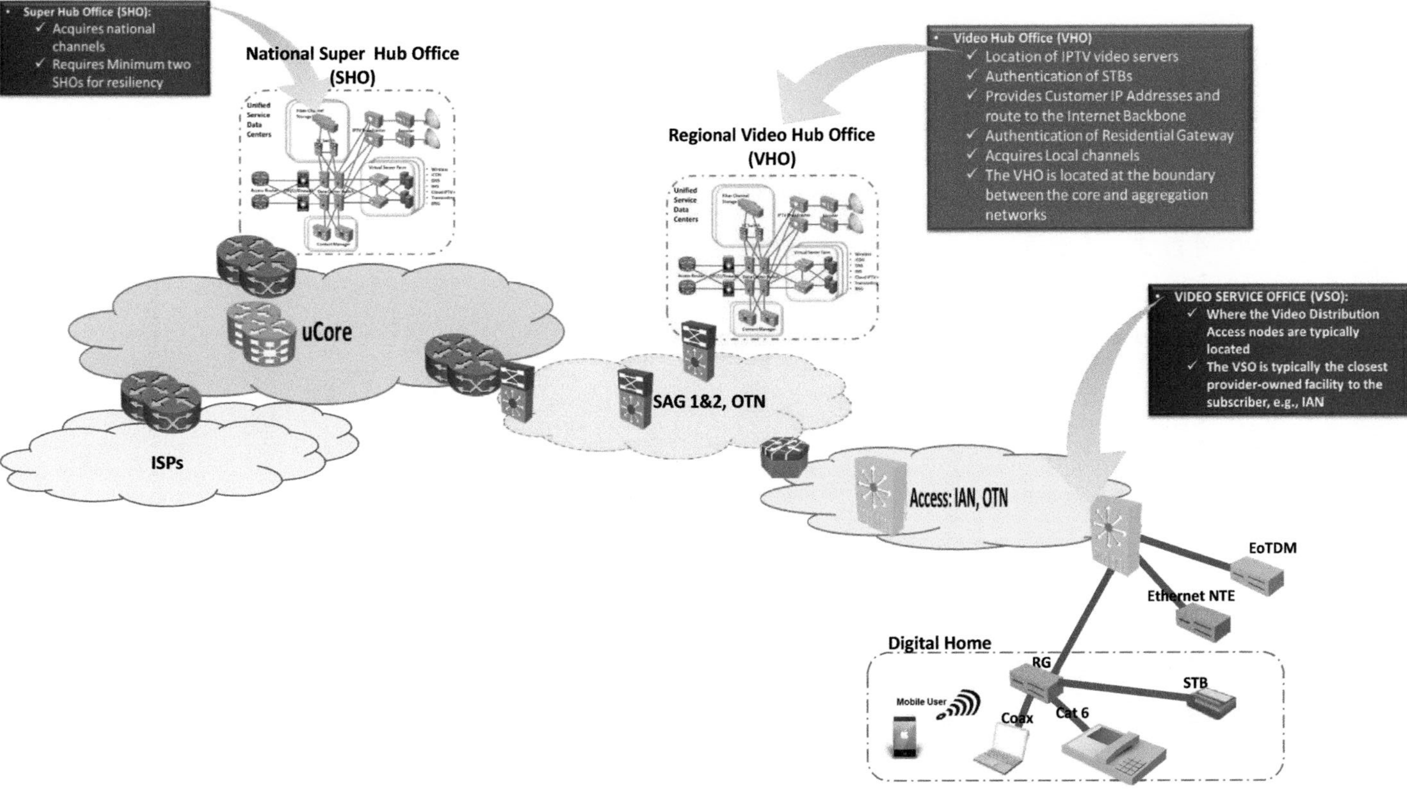

Figure 4.63. IPTV architecture.

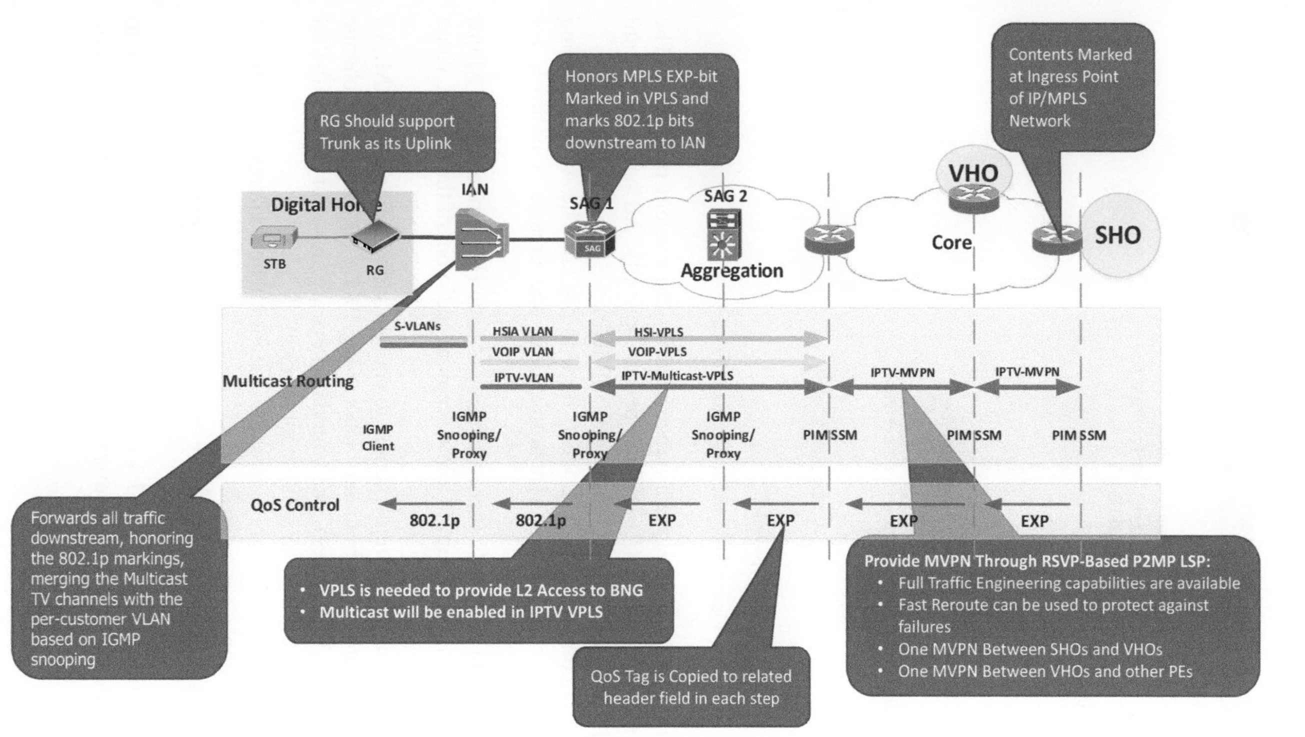

Figure 4.64. IPTV architecture.

IPTV Configuration:

IPTV configuration rules are presented in Figure 4.65.

BII Configuration Rules	Score Card
SHO:	●
• Uncompressed / lossless compression National content to encrypted MPEG2	
VHO:	●
• Receiver • Decoder • A/V Processing • Encoder • Firewall, Application servers, Client Gateway	
uAggregation: Multicast, customer VLAN	●
Access: multicast	●
In The Home: RG/MAG access to all services	●
STB/MAG: Decrypt video	●

Figure 4.65. IPTV configuration rules.

The receivers must receive signals transmitted by the content producers. The local content is received off-air (RF-NTSC) and/or optically. All national content must be received from satellite (e.g., encrypted MPEG2).

The decoders uncompress MPEG2 video back to digital. The A/V processing performs audio leveling and frame synchronization, and creates thumbnails for monitoring.

The encoders are H.264/AVC encoders. They encode HD at rates of ~ 5 Mbps, and 4K at ~ 15 Mbps. It is important to allow separate audio out of the HD/4K video stream for separate encoding.

Acquisition servers must be implemented to encrypt video and to add DRM. All content must be encrypted with boundary keys using PKI x.509 keys that change every few minutes (e.g., 30 minutes). The boundary keys are also encrypted with PKI for secure exchange and are kept in volatile memory.

The distribution Servers are key to provide instant channel change, and packet resiliency error correction.

The VoD servers must be used to store encoded and encrypted copies of VoD movies.

All IP routers must use VRRP and VPLS FRR for failover. Also, it is key to provide layer 3 access control lists with explicit permit rules (all non-conforming traffic is dropped).

IPTV Protection:

IPTV protection rules are presented in Figure 4.66

BII Protection Rules	Score Card
Min 2 SHO for resiliency	●
Resillient uCore	●
Resilient uAggragation	●
Resilient uAccess	●
MultiCast from uCore thru uAccess	●

Figure 4.66. IPTV protection rules.

4.6.3.8 Over-The-Top Streaming (OTT)

Even though OTT streaming is not a managed service, it is our collective prediction that the OTT streaming technology is an evolutionary step forward from telco provided IPTV services over time (Sujata, 2015).

The success of the OTT streaming is built upon an ecosystem where the inter-working IP networks have become much more reliable, where content is placed over a distributed architecture at the edge/aggregation relative to the eyeballs, where DRM is fully enabled, and where significant progress has been made with encoding, adaptive streaming, and compression technologies.

OTT streaming has brought a new approach, with significantly lower CAPEX unit cost and improvement in QoE, to achieve levels of streaming quality over HTTP that allows not only VoD, but also, gaming and content to be broadcast over the Internet.

OTT Streaming performance has improved significantly when it comes to the support for interactive video, such as gaming. An important QoE metric in cloud gaming is the response time, which corresponds to the elapsed time between the generation of an action, the processing, and the displayed result of that action. A game that demands low-latency, such as first-person-shooter, requires < 80 ms latency (RTT). The latency is generated by a thin client-player processing delay, the network delay, and the delay in the powerful cloud gaming application. OTT is a "best-effort Service", however, when it is built correctly, it can support VoD, interactive video, and live TV content from SD resolution to 4K and up while exceeding the QoE expectation.

There are many OTT protocols to enable digital streaming/broadcasting. The key protocols include Apple's HTTP Live Streaming, Google's WebM, Microsoft's Smooth Streaming, Adobe's HTTP Dynamic Streaming, etc. There are many key players, leveraging the above streaming technologies to offer VoD services, these include familiar names such as Netflix, Apple TV, YouTube, etc.

In the context of DSP transformation, we are not promoting one protocol/ technology over another. However, we do outline the BII capabilities that must be incorporated into an OTT streaming solution.

OTT content is not controlled, but must be protected through DRM (Bitmovin, 2018) and (encoding.com, 2020), and its QoS is not guaranteed, but must be QoE compliant through a sound architecture. However, to compensate for the nature of the "Best Effort Service", it's architecture must be uniquely developed to deliver to QoE, e.g., encoding video at different quality levels, HTTP adaptive streaming, compression, DRM, distributed architecture with multiple CDNs for content delivery, etc.

Figure 4.67 presents a high level comparison of OTT and IPTV technologies.

	Internet TV / OTT	IPTV
Transport	• Use best effort internet	• Use managed network
Geographical Reach	• Access available from anywhere in the world	• Access limited by the foot print of the service provider
Service Quality	• Best Effort Service	• Guarantees high quality audio and video
Access Mechanism	• Any device connected to internet	• Dedicated set-top-box
Content Generation	• No control on contents	• Managed by service provider
Content Control by Subscriber	• Not controlled	• Subscribers control contents

Figure 4.67. IPTV vs OTT streaming.

We recommend the following QoE expectations (Figure 4.68) for OTT services:

Macro Service Flows	QoE Expectation	Score Card
Streaming-Service flow	• Zero pixilation • Zero freeze • Sync'd voice & video	●
Interactive Video- Service flow	• Same as streaming • Action Trigger sync'd with video	●
Bandwidth-Consumer	• Guaranteed throughput (DL/UL)	●
Service Availability	• SLA compliant • High availability	●

Figure 4.68. OTT QoE expectation.

Here are the OTT network performance requirements to achieve the QoE expectations as stated above (Figure 4.69). This performance target is established to ensure QoE for one of the most demanding service flows, e.g., an interactive gaming application, e.g., FPS:

BII Performance Objective	Score Card
Interactive Video QoE	●
• Latency < 80 ms	
• Jitter < 10 ms	
• Packet Loss < 0.01%	
Bandwidth per viewing channel: 12-18 Mbps for 4K content	●
Service Availability: > 99.999%	●

Figure 4.69. OTT performance objectives.

OTT QoE centric solution must overcome many of the shortcomings being mastered by IPTV platform where the performance and the bandwidth is fully managed through guaranteed last-mile bandwidth as well as the use of MPEG short Transport Stream protocol. Here are some of the challenges for OTT to overcome: OTT is provided over the best-effort network with a wide geographical coverage over multiple cooperating ISP networks. The access mechanism is provided through any device connected to the internet at a variety of last-mile speed. Its content is not controlled by CSPs. Also, the subscriber has little to no control over the content due to the parental control capability.

Today, there are many OTT technology solutions for content delivery, however, we strongly recommend a platform solution that can handle both VoD (HTTP based format) and live/IV streaming (MPEG transport stream format).

HTTP is successfully used as a transport solution for video on demand (VOD) media embedded into web pages, e.g., on Adobe flash-based sites, such as YouTube, and Hulu. However, this solution does not stream in real-time, but instead relies on progressive downloading of media files. The browser downloads the file from the HTTP web server and when it has a sufficient amount of data, it starts to play the content while it continues to download the rest of the file. The main drawback of this approach is the length of time it takes to fill the initial buffer. Another issue associated with HTTP is streaming quality, which depends on the IP connection. Content streaming may be subject to stalling if there are fluctuations in the bandwidth, leading to frame freezing. As a consequence, it is nearly impossible to use this solution to broadcast live channels, instead, live channel broadcasting operators are using UDP multicast protocol.

To ensure the QoE for availability target, the resiliency of the network and the sources for the content are essential. For latency target, we must ensure that the provider network, as well as the interconnecting network, is multicast enabled, and protection capacity is fully deployed for the peak traffic. For the packet loss target, we must implement FRR throughout the interconnected networks.

OTT Architecture:

There are many network architectures for the OTT services, including centralized, proxy-based, distributed with CDN, etc. However, we recommend a distributed architecture as this architecture balances the QoE and high availability with the economics. It allows fast content delivery as the network consists of a multitude of edge servers/CDNs which serve the clients closest to them, while they move other services (e.g., recommendation system, content store, database solutions), such as Amazon Web Services, to the cloud for scalability.

Our recommendation for the architecture rules (Figures 4.70 and 4.71) is as outlined below.

BII Architecture Rules	Score Card
OTT Architecture: Distributed • Video encoded at various quality level, and divided in small chunks • Device Player with run-time-environment (down load, decode, and play, rank for CDN) • Cloud Content CDNs over many networks o with capability for Any-Cast to ensure load balancing and minimize server/network congestion • Cloud services (authentication, log recording, DRM, user sign-in) • OTT multi-Data centers (register users, payment)	●
Video streaming container over SSL connection (provide metadata for adaptive streaming to switch between bit rates on the basis of network performance and sensory data from the smartphone, e.g., h.264/.vmv encode)	●
Video location CDN based on popularity o with optimal update algorithm for placement of popular videos.	●

Figure 4.70. OTT architecture rules.

BII Architecture Rules- Continued	Score Card

Content viewing:
• 2 Phase streaming process (Dynamic Streaming over HTTP):
  o buffering (full bandwidth)
  o steady state (one block of data at a time)
• Download Option to allow viewing later
• UI thru HTML5 for OTT ready devices when adaptive streaming is supported
	●

Wireless client: • Native Player Application • HTML5 when supporting adaptive streaming	●
Home Client: • Open system, standard based application execution platform • Network attached Storage • DLNA compliant • Interfaces: USB, xDSL, Ethernet, VoIP, wireless • DRM to protect sharing at home	●
Common network infrastructure: • uCore • uAggregation • uAccess	●

Figure 4.71. OTT architecture rules—continued.

Overall architecture must allow encoding contents for both video and audio. It must also enable multiple subtitles and multiple languages. It is key to have the size of the video/audio small chunks in the order of single-digit seconds. The architecture also must call for a distributed HTTP server to send the content to the users in every geographical region. This architecture is only QoE effective when supported by the Adaptive Any-Cast routing capability to ensure that network congestion, as well as the CDN congestion, is avoided.

OTT Configuration:

Figure 4.72 presents the configuration rules for OTT services.

Best-in-Class Configuration Rules	Score Card
Central:	●
• Video encoded at various quality level for each region • Video distributed to the regions	
Regional:	●
• Video stored and managed for the region • Video distributed to all CDNs in the region	
CDN: • Video stored and managed for the coverage in the region • Video is served to the client in the coverage area	●
uAggregation: Multicast	●
Access: Multicast	●
Wireless client: • Native Player Application • HTML5 when supporting adaptive streaming	●
Home client: • RG/MAG access to all services • Decrypt video	●

Figure 4.72. OTT configuration rules.

OTT Protection:

Components of the OTT platform will fail from time-to-time, however, the services that are riding on this platform must not be disrupted and the QoE must not be impacted negatively.

OTT platform must be designed for failures. The boundary of the failures must be defined carefully, as it is a tradeoff between the CAPEX and QoE. Generally speaking, for a five 9's of service availability, the BII protection rules under failure are summarized below (Figure 4.73):

BII Protection Rules	Score Card
Split content by region	●
Encryption: DRM	●
Server redundancy: • CDNs for content over many networks • Cloud service (authentication, log recording, DRM, user sign-in) • OTT Data centers (register users, payment)	●
Ensure connections thru resilient: • uCores • uAggregations • uAccess	●

Figure 4.73. OTT protection rules.

4.6.3.9 SoIP Platform

IP-based services are exploding. This direction is influenced by the lifestyle change, such as blogs, social networking services, interactive video (e.g., gaming), music and video streaming, etc. As a result, it is forcing the networking technology trend towards convergence of digital services. We pioneered this global convergence to reap the benefit from the rapid introduction of new digital services, IT efficiency, and lower cost infrastructure networks.

SoIP (Service Over IP) is a transformational platform for the DSPs to design and to create digital services to connect people globally, providing meaningful content and applications, enabling decision making, and allowing people to work and play seamlessly anywhere, with any device, and at any time (at&t, 2010).

These services must be provided seamlessly, independent of the access technologies of wireless, cable, wireline, power line, Wi-Max, etc. Also, it must be provided transparently when switching over between the access and device technologies. It is also critical to establish a shared architecture for consumer and business services, i.e., it must support multiple, concurrent, user personas, and be able to concurrently maintain both business and consumer sessions.

The unified SoIP infrastructure must enable: (1) being used by the internal users (e.g., employees), (2) used by the DSPs for their customers, and (3) offered as service to the competitive service providers- the latter functionality is instrumental in mergers and acquisitions.

The overall architecture for SoIP is comprised of four components (Figure 4.74): (1) a unified Service and Design Execution Environment (includes application servers), (2) an IP/MPLS converged network, (3) service specific border controller, and (4) Device Specific Execution Run-Time Environment.

In this architecture, each access technology communicates via a specific border controller which provides a uniform internal view while addressing the uniqueness of the external interfaces. To provide a new service, the corresponding application server is "plugged-in". Once a new access technology is supported, it can be used

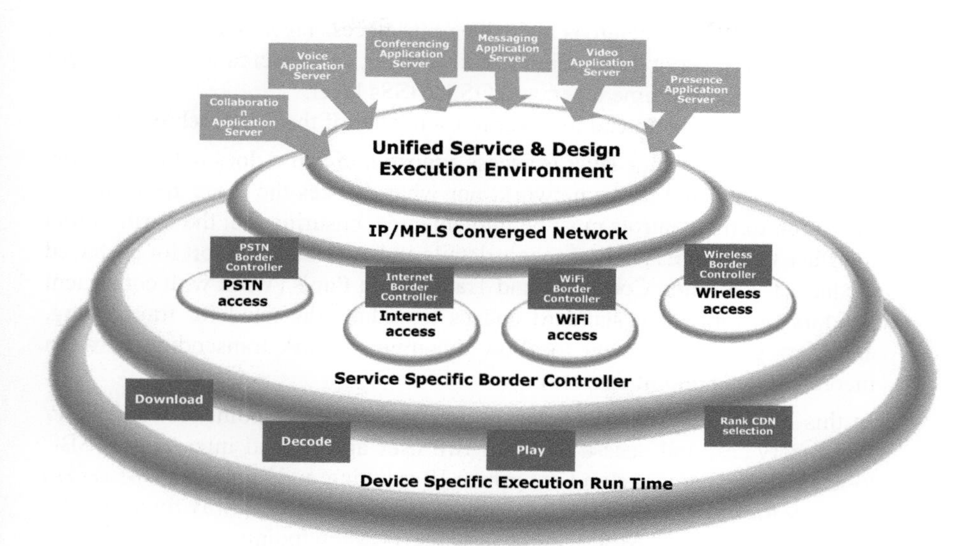

Figure 4.74. SoIP platform framework.

by all existing and future application servers, and once a new application server is deployed, it can support all existing and future access technologies.

Here is the overall architecture (Figure 4.75) for service realization and execution for DSPs:

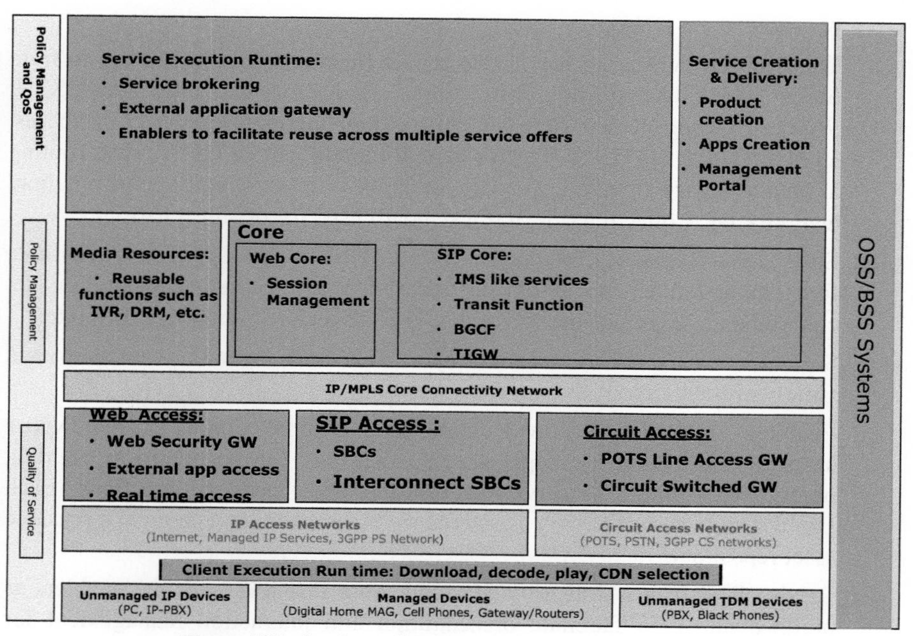

Figure 4.75. Service realization architecture—conceptual view.

The overall SoIP architecture is inclusive of the access layer, network core layer, media resource layer, service execution run-time layer, service creation and delivery layer, client execution run time layer, and OSSs/BSSs layer.

Here is what we recommend to cover in each layer of the SoIP architecture:

(1) SoIP Access Layer: this layer is access agnostic, in that it does not matter how users are connected to the network, nor what devices the users are using. At this layer, traffic controls must be implemented, ensuring that the north-bound interface to the core is hiding access details and providing support for preferred Codec set- Common Codec Set and Transcoding Pools (Work with equipment vendors to implement standard codecs to reduce the need for transcoding, and New network element to allow dynamic network transcoding between incompatible endpoint devices).

In this layer, you will need a security GW as an entry point to web/HTTP/Portal services, and access SBC for SIP user agents, and interconnect SBC to access peer networks and enterprise IP trunking endpoints, a TDM access gateway for individual POTS lines, and a circuit switch gateway for access to/from PSTN, wireless MSCs, and enterprise TDM endpoints.

(2) SoIP Core layer: In this layer we must provide Web core services, as well as SIP core. Web core services include web session management, where the user could invoke and manage multiple web/HTTP services with the same dynamic collaborative session. SIP core services provide "IMS" like services, including I/S- CSCF (call signaling protocol), HSS (master database for a given user), and ENUM (mapping a telephone number to a URI specifying a host that can be resolved via DNS).

The SIP core must also handle the transit function (directly routing to/from border elements bypassing IMS core), BGCF (it is the intelligent PSTN/Wireless routing element for LNP, toll-free number translation, N11 databases, and HLR), TIGW (TCAP IP gateway is for intelligent PSTN/Wireless routing as well as legacy signaling gateway for application servers), Support function, and the CDR Collection (the 3GPP standard used for off-line charging).

(3) SoIP Media Resources: Media servers are to be designed to provide reusable functionality, independent of the underlying services such as multimedia conferencing, announcement services, IVR, speech recognition, DRM, etc.

(4) SoIP Service Execution run-time layer: This SoIP layer must encompass a variety of functionalities:

Enablers—a set of well defined, autonomous, runtime components that expose an open interface to the application so that they can be reused as building blocks across multiple service offers and other network functions. Each Enabler must be certified by a governance process and registered in a reusable asset repository.

Service Brokering—a distributed architecture that leverages the enablers as outlined above to facilitate orchestration and interaction management for various types of applications provided by their offering as well as the other service providers, including SIP and Web Services. Capabilities must include

Feature Interaction Management capabilities and the ability to synthesize and execute complex services from simpler constituent parts.

External Application Gateways—exposes network capabilities to DSPs' partners and customer applications. The EAG manages bidirectional access between external applications and key DSP capabilities, supports SIP and IMS through a variety of service delivery models, and supports the interaction of Enablers with external application providers.

(5) SoIP Client Runtime Execution Environment—a computing infrastructure for high performance, high availability and scalability suitable for hosting runtime service components.

At this layer, the runtime environment is capable of downloading content, decryption, play, and rank for CDN selection all in real-time. The common shared platform provided by REE supports converged application services and incorporates support for external 3rd party apps, partners and customer environments.

(6) OSSs/BSSs interface—the key function for the OSSs/BSSs is the ability to connect to any layer and obtain information to deliver the services and render accurate billing and perform service assurance activities. These API-based interfaces must allow for real-time provisioning, real-time charging, SLA management, and real-time access management to enable user authorization and management.

(7) SoIP Service Creation and Delivery (SC&D) layer: This SoIP layer must address the ecosystem for application creators, advertisers, service providers, and end-users, and must encompass the following functionalities:

Application Architecture—introduces a component model and service delivery framework for Applications.

Encapsulates OSS/BSS variation and complexity behind a required Web X.0 component presentation layer.

Enables a Services Marketplace in which DSP acts as a broker between third party component creators and third-party service providers while encouraging resale of DSPs' OSSs/BSSs capabilities.

Product/Offer Creation Environment—portal and wizard enabling internal and third-party product managers to configure products and offers from DSPs' application components registered with the DSP Services Marketplace.

Application Creation Environment—SDKs and libraries facilitating the creation of application components. Libraries ease the creation of the application presentation layers. SDKs provide access to enabler capabilities (via enabler APIs) OSSs/BSSs functions.

Management Portal—Architecture supporting the ordering, monitoring and management of offers from a variety of devices and existing portals.

(8) OSSs/BSSs Layer—the following functions must be integrated into the OSS/ BSS Architecture:

Policy Management

- Dynamic workflow and Resource Allocation
- Real-time Inventory and Data Management
- Portal

SC&D Realization Via Product Catalog Infrastructure

- Product Catalog Repository
- Marketing Tools, Product Management Tools, Service Definition Tools, Service Guide

3rd Party Provider Management Tools: On-boarding, Billing Settlement, Account Management, Assurance

Sales & Service Fulfillment Enhancements

- Converged Sales Model use single customer instance across lines of business
- Persona Management
- Intelligence-Based Proactive Sales & Provisioning
- Integrated Ordering
- Multi-domain Order Distribution & Management
- On-demand Service Provisioning
- Dynamic Workflow for Service Activation
- Common support functions across wireline and wireless domains: E911, CALEA, billing, etc.

Service Assurance Enhancements

- Enhanced temporal, spatial and topological Correlation
- Move to granular real-time Fault and Performance Monitoring
- Implement Advanced Active & Passive Monitoring
- Dynamic SLA/QoS Management
- Dynamic real-time management of the SoIP environment
- Scalable, highly available services
- Integrated Policy/SLA Management
- Charging & Billing Functions
- Persona-based Billing
- Online/Offline Charging
- Credit Control & Real-time Balance Management
- Charging Rules integrated as Policy
- QoS or Event-Based Adjustments
- Intelligent Usage Collection & Analysis

Bottom line, BIISoIP must enable DSPs to offer both legacy services and new converged services within the following construct: access agnostic (wireless, wireline, cable, WiFi) converged user devices, concurrent user personas, and global mobility where same services can be provided everywhere, internationally, at home, in the car, at the office, etc., this capability is best placed at the edge of the IP/MPLS network.

4.6.3.10 Unified Data Centers

Legacy CSPs' POPs and data centers are built and operated in silos across line of businesses (wireless/wireline/video) and house specialized hardware devices that are costly and require extensive management operations to keep them running and maintained at the prescribed performance level. Also, the poor placement of these legacy data centers often impedes the QoE. These legacy complexes are housed redundantly in 6 distinct types of POPs/Data Centers. These Complexes are comprised of National/Regional POPs, Metro POPs, Local Distribution/Cell site POPs, service-specific data centers (at national and regional levels), and IXPs. Besides, the legacy POPs and data centers are built redundantly, in silos, to support each wireless/wireline/video infrastructure. For example, there are 3 access/aggregation/core POPs types for each service, such as wireline, wireless, or video.

Furthermore, the legacy POPs and the data centers were always separated into a hybrid complex due to the unique character of their functionalities for transport and routing vs services platforms.

The legacy data centers and the POPs (for wireless, wireline, video) are counted overall in high 10,000's for a tier 1 global CSP (including the CDN nodes, cell sites, and legacy POPs). These legacy complexes must be transformed, i.e., consolidated, standardized (modularized), automated (e.g., provisioning, configuration), virtualized and simplified into low 10,000's of Unified Data Centers and POPs—a minimum of 5 to 1 reduction.

The FMO Unified Data Centers must provide two distinct functions: (1) Unified Infrastructure Data Centers (UIDC, also known as POPs) providing specific transport functionality including access, aggregation and routing for one unified network, and (2) Unified Services Data Centers (USDC) providing application-specific functionality, such as SoIP, gaming,VOD, etc., for one unified network. Here is the functionality of the transformed data centers.

(1) Unified Infrastructure Data Center (UIDC): Unified Infrastructure Data Centers must be distributed throughout the DSPs' footprint at the National/Regional, Metro, and Local Distribution level. The only function of UIDCs is to provide access, packet collection, grooming, aggregation, routing and transport, and in some cases interconnect functionality with the cooperating ISPs. It is important to note that some select UIDCs may be connected to or housing USDC to provide service functionalities for the end-users as well:

    i. National/Regional UIDC: these data centers (counted in 10's) are distributed throughout the uCore, providing inter-regional transport and routing, and must contain platform infrastructure for integrated wireless/wireline/video packet aggregation, grooming, label processing, and backbone routing.

    ii. Integrated Metro UIDC: these data centers (counted in low 1000's for a tier 1 DSP) are distributed throughout the uAggregation, providing depth and reach to quickly and efficiently process and appropriately transport packet traffic, and must contain platform infrastructure for integrated wireless/wireline/video packet aggregation, grooming, label processing, backhaul/transport, and access protocol.

iii. Local Distribution UIDC: these data centers (counted in high 1000's) are distributed throughout the last mile, providing a finer level of depth and reach to provide last-mile connectivity, and quickly and efficiently collect, process and appropriately transport packet traffic, and must contain platform infrastructure for access protocol and packet aggregation.

iv. IXP: this functionality is outlined under the "Unified Support Infrastructure".

(2) Unified Services Data Centers (USDC): USDCs are distributed selectively throughout the DSPs' footprint at the national/regional level, metropolitan level, as well as placed in the cooperating ISP networks at the global level. The main function of USDCs is to provide SoIP and network-specific functions, such as caching/streaming, NFV and SDN control plane, as well as the OSSs and BSSs to run and operate the DSP infrastructure. The USDCs must be placed strategically, e.g., at the edge of the CSP's network (to ensure QoE for the on-net eyeballs and IoT applications) (Sowry, 2018), or in the cloud at select global footprints (to provide economy of scale for cloud services for external/internal customers), or at the data centers distributed globally at the select data centers operated by the cooperating ISP providers (to ensure QoE for the off-net eyeballs and IoT applications).

USDCs are comprised of the Edge DCs (connected to DSPs' network), Cloud DCs (connected to DSPs' global network), and iCDN DCs (connected to DSPs' as well as cooperating ISP networks). It is important to note that some select UIDCs may be connected to or maybe housing USDC to provide service functionalities for the end-users:

i. Cloud computing Data Centers: these data centers (counted in 10's) are distributed throughout the global/national network, connected to the select PEs, providing application-specific services, and must contain platform infrastructure for an integrated wireless/wireline/video service infrastructure, such as sign-in/registration, payment, authentication, IMS, LTE MME, NFV, SDN, IPTV, etc.

ii. Edge/iCDN data centers: these data centers (counted in 1000's) are distributed throughout the DSP's footprint, connected to the aggregation/PE nodes and/or connected to cooperating ISPs POPs, providing streaming, and caching functions and the infrastructure.

Unified Infrastructure Data Centers:

UIDCs provides almost the same functionality throughout the DSP network in the aggregation/grooming/routing functionality. However, the key differentiator is the scale at which this functionality is taking place. Here is a high-level design objectives for these data centers:
Figure 4.76: Performance objectives

BII Data Center Performance Objectives	Score Card
QoE Performance (PE-to-Data Center): • Latency < .4 ms (0.1 ms per routing/transport element) • Jitter: Zero • Packet Loss: Zero	●

Figure 4.76. UIDC performance objectives.

Figure 4.77: Architectural Rules

BII Architecture Rules	Score Card
Distance from PE < 80 Kilometers	●
PE-to-PE/P Latency < 12 ms per 1000 Kilometer	●
Dual platform for PE/P	●

Figure 4.77. UIDC architecture rules.

Figure 4.78: Protection Rules

BII Protection Rules	Score Card
Data Center Resiliency Failover Mode	●
• One Data Center	
• One Fiber Span	
• Two Fiber Span	

Figure 4.78. UIDC protection rules.

Figure 4.79: Configuration Rules

BII Configuration Rules	Score Card
Connectivity PE-to-PE, P-to-P at link speed of 100+ Gbps	●
Dual platform for Routing	●
Quad switching platform	●

Figure 4.79. UIDC configuration rules.

(1) Integrated National/Regional UIDC: providing inter-regional transport and routing, and must contain platform infrastructure for an integrated wireless/wireline/video packet aggregation, grooming, label processing, backbone routing. The Integrated National/Regional UIDC, in some cases, will house or will be connected to USDC for application infrastructure through the PE node. With this arrangement, integrated applications (e.g., SoIP), and specific applications (e.g., LTE MME) that are all IP reachable, will be provided through the uCore for consistent and expanded reach across the DSPs' footprint.

(a) Internet connectivity is provided through a unified core, consistently and optimally across wireline, wireless, and video,gaining economies through common MPLS network and security policies and management.

Also, it provides for tighter integration and QoS with a wireline Broadband access network to deliver at scale, applications such as WiFi offload/small cell/5G+ providing for the lowest cost/bit transport and furthering opportunities for advanced applications across wireline/wireless/video through SoIP.

(b) Point-to-point connectivity at the High-speed of 10Gbps+ is provided through OTN infrastructure.

Here is a pictorial representation of an Integrated "P"/"PE"/GW (Figures 4.80, 4.81, and 4.82) UIDC to cover the aggregation functions:

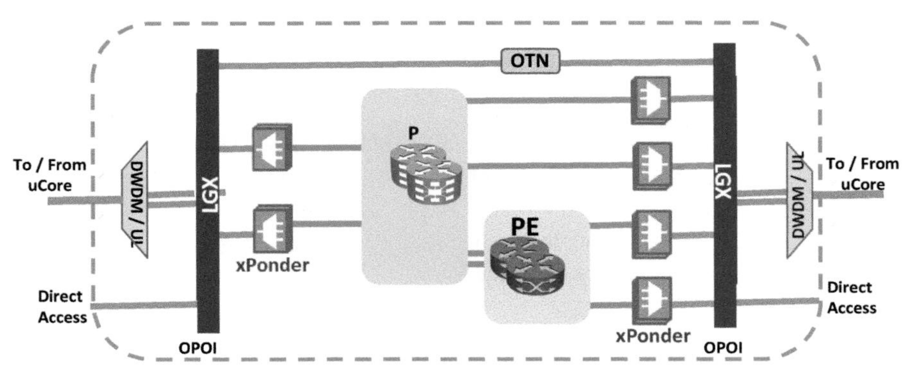

Figure 4.80. Unified infrastructure data center (UIDC) core node standard architecture.

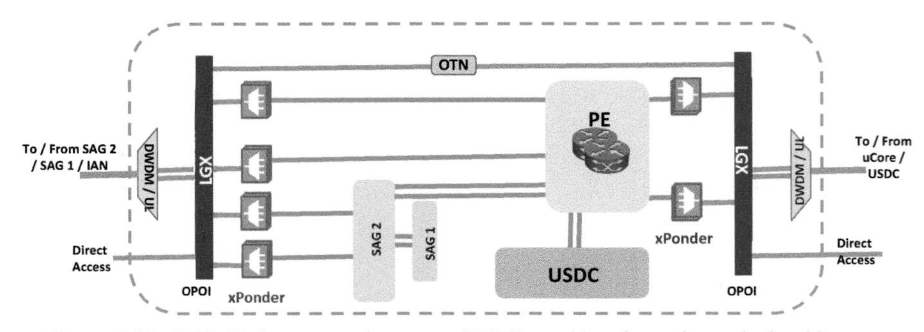

Figure 4.81. Unified infrastructure data center (UIDC) provider edge node standard architecture.

(2) Integrated Aggregation UIDC must house all infrastructure, providing depth and reach to quickly and efficiently aggregate and appropriately transport packet traffic. The Integrated Metro data centers, for example, will have network gears for access/aggregation, wireless SGSN, RNC, LTE aGW and 3G RNC and SIAD/MSN integration, various circuit switch (providing a circuit to packet) gateways and transcoding pools controlled by SoIP and transit functions.

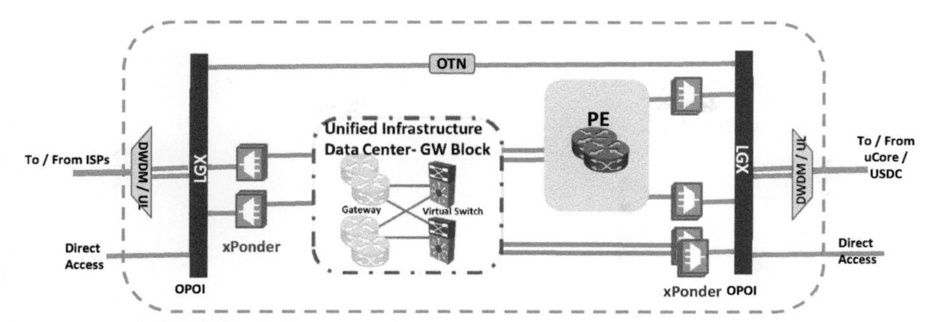

Figure 4.82. Unified infrastructure data center (UIDC) gateway node standard architecture.

As MPLS functionality is expanded into the uSAG, USDC can be directly connected to the uSAG nodes to provide application-specific services to the end-users without having to traverse through the uCore/regional core, this capability will be instrumental in improving QoE as well as the resiliency of the service containers. The uSAG extends into the integrated Metro POPs and provides not only backhaul from the Cell Site's but also assumes all of the L2 and logical layer functionality and connectivity required by wireless for interconnection amongst devices as well as interconnection to the uCore. This allows (for example):

- ✓ wireless and wireline subscribers to both reap the benefits of CDN (content delivery networks) as those are extended across the footprint.
- ✓ High bandwidth local IP LTE bearer traffic can be turned around locally at the aGW rather than traverse the uCore network (QoS, transport efficiencies).
- ✓ Additional design and planning flexibility by allowing for co-located mobility equipment with wireline equipment to maximize scale and economic reach and reduce backhaul.
- ✓ Significant real estate savings and transport economies due to the co-location of wireless/wireline/video equipment.

Here is a pictorial representation of an Integrated Metro UIDC to cover the aggregation function (Figure 4.83):

(3) Integrated Local Distribution UIDC must house all infrastructure, providing finer depth and reach to quickly and efficiently aggregate and appropriately transport packet traffic. The Integrated Local Distribution UIDC, for example, will have network gears for access/aggregation, and transport to the Aggregation POPs.

Here is a pictorial representation of the Integrated Local Distribution POP (Figure 4.84):

Disaster Recovery:

The disaster recovery for UIDC must include the node architecture for "P" node, "PE" node, uSAG node, uAccess node (e.g., IAN), IXP node, GW node, and the optical node, as described in the associated sections above.

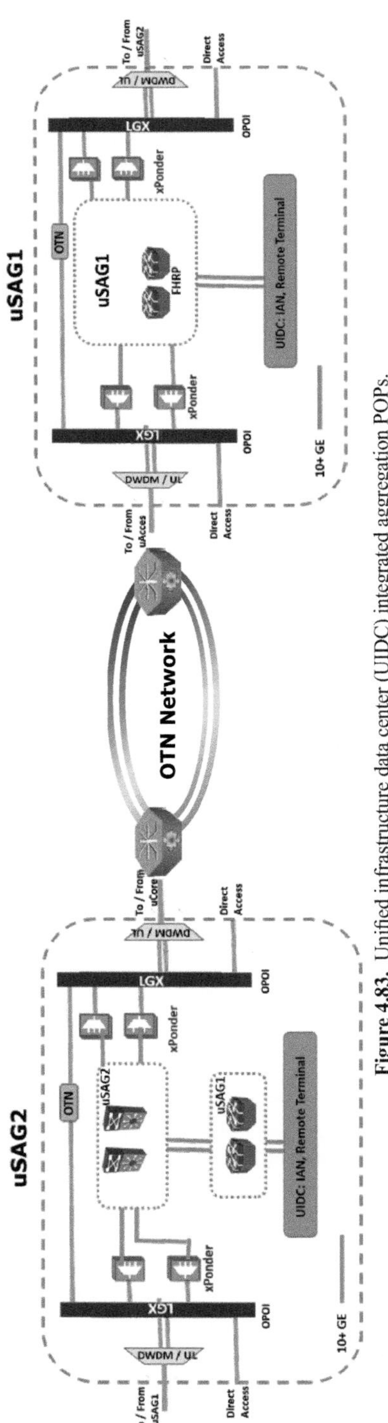

Figure 4.83. Unified infrastructure data center (UIDC) integrated aggregation POPs.

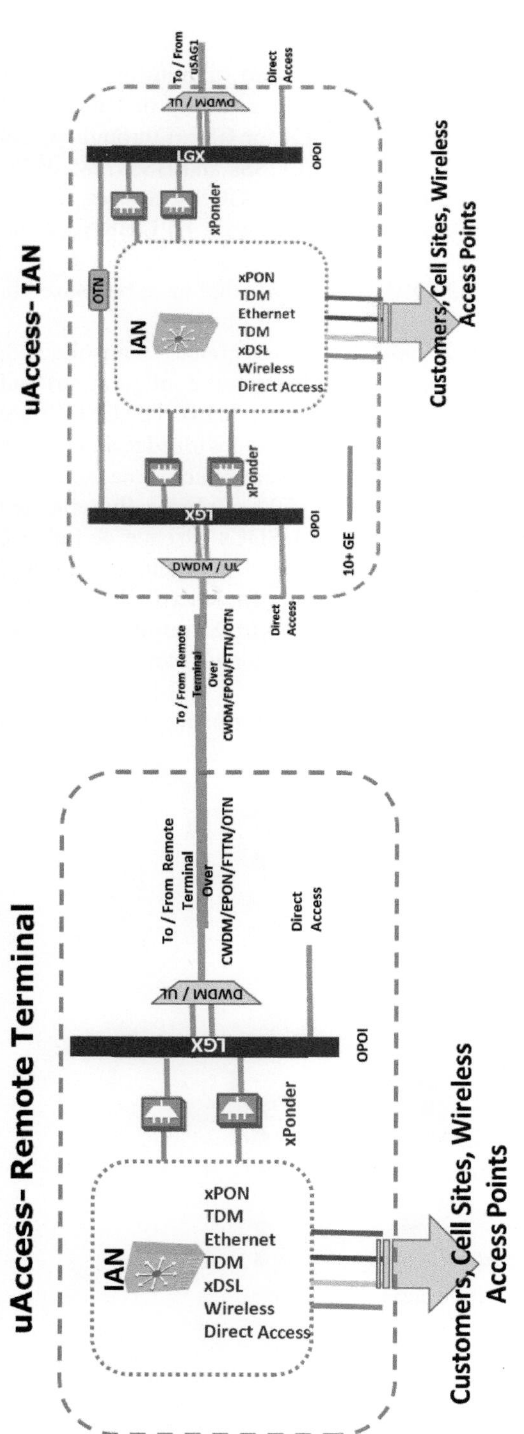

Figure 4.84. Unified infrastructure data center (UIDC) access network.

Unified Services Data Centers:

There is a plethora of application-specific systems that are running on a mix of physical and virtual machines and network equipment in a variety of legacy data centers (IDCs for end-customers as well as Private DCs for CSPs) throughout the CSPs' footprint of wireline/wireless/video, e.g., silos of OSSs and BSSs, IM, IMS, SoIP, SMS, DNS, RADIUS, MIND, LTE aGW, LTE MMEs, GGSN, Session Border Controllers, Femto GW, 3G SGSN, SHO, VHO, DNS, Firewall, DPI, BRAS, Colo services, managed services, and cloud services.

There are also SDN control plane and NFV technologies that must be housed in the data centers.

These application-specific systems must be integrated, consolidated, standardized (modularized), and automated (e.g., provisioning, configuration) and run in Virtualized Data Centers that we call USDCs (Sowry, 2018). The USDCs are best to be categorized in two types for a broad base deployment with edge and remote cloud data centers, where edge data centers will complement remote data centers in providing higher quality online services at low cost: (1) 10's of physically separated Virtualized (Software Defined) Cloud Data Centers—USDCs (Private & Public), and 2) 1000's of Virtualized (Software Defined)Edge/iCDN Data Centers—USDCs. It is important to note that a Software Defined Data Center could easily house many virtual data centers, each equipped with programmable virtual resources such as CPU (dedicated to run applications and the operating systems), storage, memory, and networking (such as switches, routers, and the links).

We recommend the following QoE driven network performance for the USDCs (Figure 4.85):

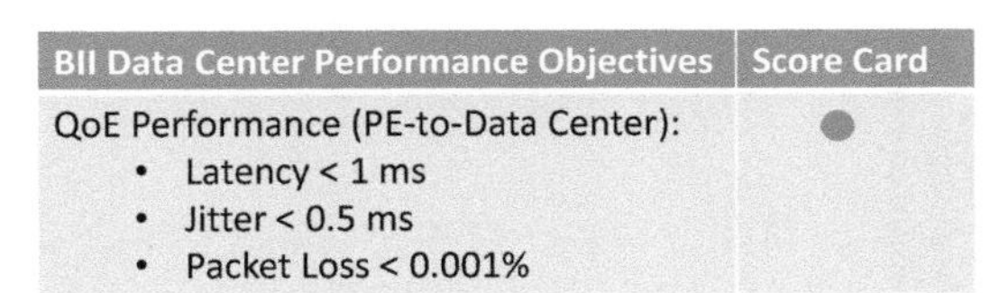

BII Data Center Performance Objectives	Score Card
QoE Performance (PE-to-Data Center): • Latency < 1 ms • Jitter < 0.5 ms • Packet Loss < 0.001%	●

Figure 4.85. USDC performance objectives.

Architecture rules (Figure 4.86):

BII Architecture Rules	Score Card
Distance from PE < 80 Kilometers	●
PE-to-Data Centers Latency < 12 ms per 1000 Kilometer	●
Dual platform for DPI/LI/Firewall	●
Quad switching platform	●

Figure 4.86. USDC architecture rules.

Here is the configuration (Figure 4.87):

BII Configuration Rules	Score Card
Connectivity to the PE at link speed of 10+ Gbps	●
Dual platform for DPI/LI/Firewall	●
Quad switching platform	●
Server farm	●
Storage farm	●

Figure 4.87. USDC configuration rules.

Protection rules (Figure 4.88):

Best-in-Class Protection Rules	Score Card
Data Center Resiliency Failover Mode	●
• One Data Center	
• One Fiber Span	
• Two Fiber Span	
Build Any Cast	●
Build Data Center Capacity/connectivity for the Peak Traffic, i.e., 95 percentile	●

Figure 4.88. USDC protection rules.

Unified Service Data Center—Cloud Computing- Public/Private:

Cloud infrastructure, both public and private cloud, must offer best-in-class services across Software as a Service (SaaS), Platform as a Service (PaaS), Infrastructure as a Service (IaaS), Data as a Service (DaaS), and private infrastructure as service, while ensuring security and 99.999% availability of the Cloud Services.

Most cloud computing services must be provided self-service and on-demand, so even vast amounts of computing resources can be provisioned in minutes, typically with just a few mouse clicks—giving businesses greater flexibility and taking the pressure off of capacity planning.

Here is an overall architecture for a Cloud-based data center (Figure 4.89). The foundation of this architecture is focused on disaggregation of functionalities that used to be running all on a single processor (e.g., x86). The Software Defined Data Center architecture is based on virtual resources, i.e., programmable virtual servers (CPU, storage, Memory), and virtual networking (switches, routers, and the links). This architecture is scalable (from a small iCDN node of single-digit Gbps throughput, to a highly expanded virtualized data center measured in Tbps), and in compliance with the architectural rules outlined above.

Unified Service Data Center- Edge/iCDN:

One of the key core competencies of the DSP must be its QoE driven digital services. The digital services are, generally speaking, comprised of caching services (web acceleration and file download) and streaming services (live webcasting and video on demand). The placement of the digital services in the global/national DSP footprint

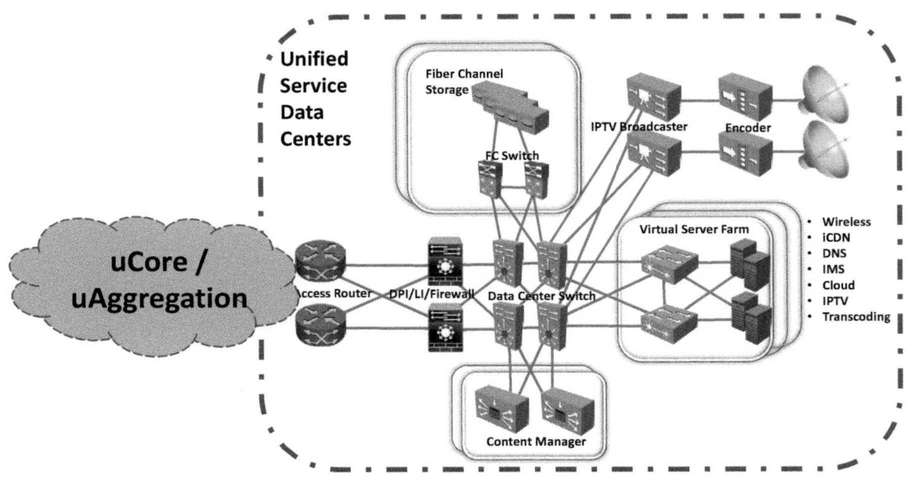

Figure 4.89. Unified service data centers.

is key in ensuring competitive QoE performance. In that, these digital services, e.g., bandwidth-intensive, rich multimedia content, such as gaming, e-commerce transact, news and entertainment services, must be placed, ideally, at the edge of the network(Sowry, 2018), or one hop away from the on-net and the off-net eyeballs to be viable from a QoE perspective (Frank, 2013).

We highly recommend a hybrid CDN architecture (Akamai, 2020), i.e., the DSP must own and operate an overlay of CDN data centers throughout its footprint, and leverage the network of other ISPs to expand the coverage of CDN throughout the ISP networks.

However, the BII CDN is assured by IP Anycast (Flavel, 2015), that allows multiple instances of the same service to be "naturally" discovered, and requests for this service to be delivered to the closest instance. This technology is further enhanced by implementing a load-aware IP Anycast CDN architecture. This architecture makes use of route control mechanisms to take server and network load into account to realize load-aware Anycast. The routing architecture through iDNS have already improved content routing by close to 500 route miles on international core, and significant improvement in QoE. However, adaptive Anycast has further reduced on-net mileage by another 50% and improved the stability of QoE.

Another key consideration for the iCDN is node placement. iCDN nodes, once placed properly, given the traffic for the on-net vs off-net eyeballs, will improve QoE performance and the network unit cost. iCDN locations must be optimized as close to the destination eyeballs as possible, this way, it will consume less network capacity, fewer potential congestion points, and have much lower latency, and of course better QoE performance.

For off-net traffic, the most cost-effective delivery of content is to place it right behind the hybrid CDN/Peering routers. In our practices, this approach resulted in a 50% reduction of the port cost, and backbone transport. Given the BII substantiated facts and experience, any content with > 60% off-net eyeballs must be placed at the CDN/Peering router locations.

For on-net traffic, the most cost-effective delivery of content is to place the content at the service complexes attached to the PEs. In some carefully measured cases, the iCDN could be placed in the uSAG network.

The below architecture (Figure 4.90) is based on high-speed uplink to the uCore (range of 10 through 100's Gbps); it must be supported with intelligent DNS and Anycast. The server virtualization must enable scalable and on-the-fly resource allocation, with availability for live VM migration.

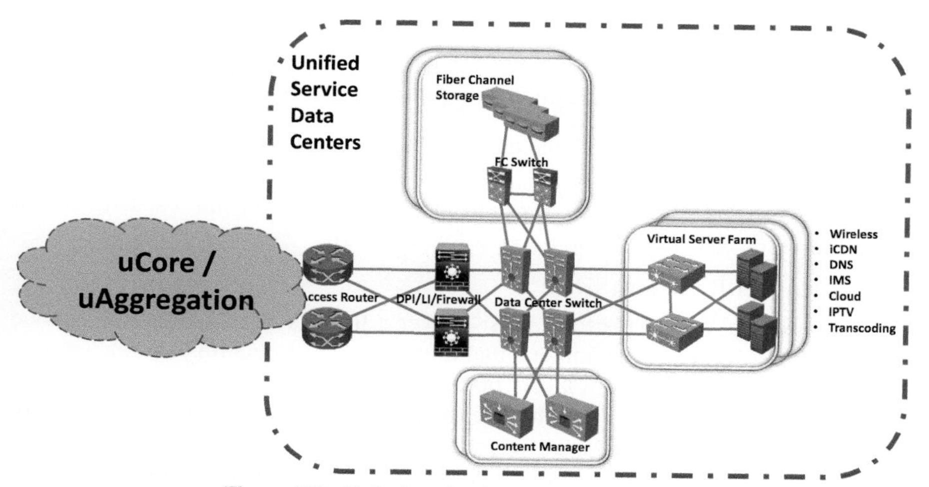

Figure 4.90. Unified service data center—edge/iCDN.

4.6.3.11 Unified Support Infrastructure

The support infrastructure for DSPs is inclusive of Network Interconnect, Disaster Recovery, and Network Security.

Traditional support infrastructure (for wireless, wireline, video, data, and content) must be transformed into unified service support infrastructure to ensure sustainable and unified interoperability with other networks, as well as full recovery from a disaster, independent of the underlining services. Unified support infrastructure is aimed at the integration of such infrastructures across wireless/wireline/video services.

Network Interconnect:

The network interconnect can be achieved through a variety of ways, including transit, peering, and IXP. The transit approach requires cost associated with the circuits and data volume. The peering approach, on the other hand, does not require cost associated with the data volume, but requires cost associated with the circuits. It is scalable with dedicated P2P transport. Lastly, the IXP approach, where all parties share the transmission equipment cost and it is scalable through the establishment of transport to the IXP for the interested party. The BII practice is to leverage the IXP approach as much as possible. This approach is scalable and least costly.

The picture below (Figure 4.91) shows the standard architecture for a gateway structure through which to connect to the IXPs. In this architecture the provider's

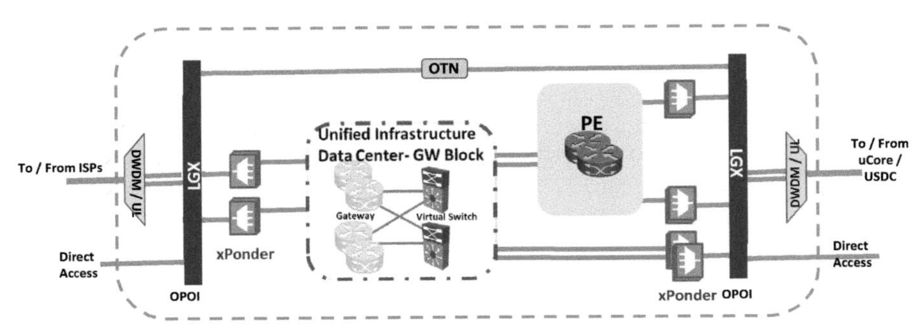

Figure 4.91. Unified infrastructure data center (UIDC) gateway node standard architecture.

routers must connect to the IXP redundant switches and must use eBGP for peering purposes.

It is important to enforce regulation to ensure minimum link capacity for all providers to mitigate poor service access for small ISPs.

Service Disaster Recovery (SDR):

As stated before, a DSP platform must deliver a five 9's of reliability and service availability. We strongly recommend that the network architecture be designed with redundancies in each of the 6 layers of the architecture in order to ensure the five 9's of reliability: (a) Transport, (b) Access, (c) Aggregation, (d) routing, (e) SoIP (Service over IP), and (f) Disaster Recovery.

Given that we have designed resiliency into the FMO network at the five 9's of availability, there are still unpredictable disasters that can happen—when they do, CSP must have the capability to tackle these by triggering a BII Disaster Recovery capability (Wrobel and Wrobel, 2009)—through a BII fleet of recovery trailers and a well-established DR process and well-practiced response team that can react quickly anywhere across the globe to restore services within max 72 hours.

Disaster recovery must be enabled by standard node complexes with which to recover from any catastrophic failures. These node complexes are comprised of data nodes, service nodes, transport nodes, power and cooling nodes, etc.

It is also worth noting that the CSP's OSSs infrastructure must also be developed to be resilient to an outage with multiple redundant IT "Class A" data centers, housing the strategic IT applications, with service recovery centers defined and on standby.

The rules for disaster recovery are outlined below:

I. DR Performance Objectives (Figure 4.92)

Best-in-Industry Network Disaster Recovery Performance Objective	Score Card
Service Recovery: • 7X24 Monitoring • Service Recovery < 72 hours	●

Figure 4.92. Disaster recovery performance objectives.

II. DR Process Rules(Figures 4.93 and 4.94)

Best-in-Industry DR Process Rules	Score Card
Duty Officer: • 7X24 Monitoring • Responsible for DR activation • Use of Command, Control, and Communication Process	●
Real time access to UIDC/USDC profile	●
Assignment and deployment responsibility: • DR team • Dr Trailers	●
Verification Responsibility: • NOC • DR team • Duty Officer	●

Figure 4.93. Disaster recovery process rules.

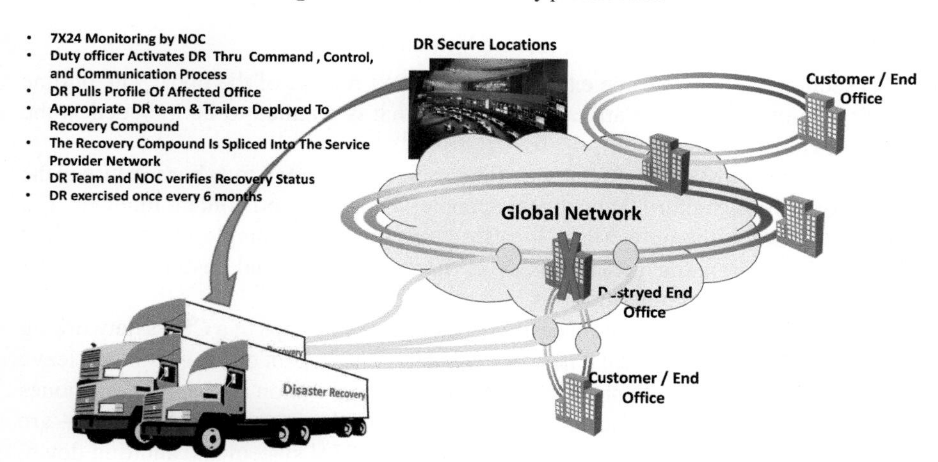

Figure 4.94. Disaster recovery action.

III. DR Configuration (Figure 4.95)

Best-in-Industry DR Configuration Rules	Score Card
UIDC trailers: • uCore, uAggregation, uAccess, IXP, uTransport	●
USDC Trailers: • Virtual Servers: SoIP, IPTV, NFVs • OSSs, BSSs	●

Figure 4.95. Disaster recovery configuration.

IV. DR Protection (Figure 4.96)

Best-in-Industry DR Protection Rules	Score Card
Multiple DR complexes: • Geographically dispersed • Each DR complex capable for full recovery from catastrophic events	●
DR Exercise every 6 months	●

Figure 4.96. Disaster recovery protection.

The BII practices suggest that all UIDCs and USDCs are expected to be standardized, scalable, and built to a set specification. These data centers must be profiled and databased throughout the DSP network. As a part of the SDR, it is expected to build a series of special-purpose trailers (e.g., transport, routing, power, etc.) to match the most scaled up nodes (data centers) in the DSP network. Please refer to the disaster recovery section of every component of the network to establish the scope for the disaster recovery standard node infrastructure.

Unified Network Security (future release Volume II):

[This section is planned to be expanded in the next release of the book. The outline below is aimed to provide an introduction to what is planned to be included in the next release.]

One of the most damaging targets for a society embroiled in cyber warfare is the networking infrastructure that connects people, machines, and content together. The users' reliance on the networking infrastructure focuses on single points of failure that can have dramatic consequences if directed at financial institutions, power stations, transport and other utilities.

A simple DDoS attack could easily wipe out 100's of Tbps of a CSP's networking capacity, i.e., > 30% of the total network capacity. The impact could very much leave millions of CSP users with slow to no service for the duration of the attack. iPhones and any smartphones or other devices connected to the internet—even IPTV—are vulnerable. Can you imagine hackers attacking the IPTV system and shutting down, for example, the Super Bowl broadcast? It's possible!

Today, each tier 1 CSP is fending off millions of attacks every day. This number is increasing at an alarming rate as billions of IoT devices are being added to the networks.

When it comes to security, you can never eliminate the risk, only minimize and mitigate it—there is no such thing as a 100% secure system.

ETE cybersecurity is comprised of security apparatus for the users, access, aggregation, core, content, and the inter-connections. The cybersecurity can be engineered and built for a closed network, or an open internet network.

We need to design a network that is cloaked. If we can make an aircraft invisible by using certain materials to evade radar, then why shouldn't we focus on cloaking infected computers, ensuring that they cannot affect other machines on the internet. Current security technologies are all after-the-fact, but we must build a network where security will follow a 3P Model:

Proactive Cybersecurity—Systems proactively and in real-time scan for flaws and use ACL to control access to software and systems in both enterprise and tier-1 service providers.

Preventative Cybersecurity—Networks adopt a set of rules to help prevent any attack on any element connected to the internet. Stochastic processes will also be a great help. Intrusion Detection Systems (IDS) are practically useless, and companies sell these technologies even though they offer no help.

Predictive Cybersecurity—Systems use real-time data and deep-packet inspection to look for patterns of attacks in order to get into the minds of hackers and stop them before they launch. A network that is capable of cloaking itself must be designed. Government assets do need to use this technique to prepare for future cyber-attacks, an effort that will require federal legislation. This is the only way to avoid major cyber wars in the 21st century. If not addressed, the negative impact of cyber wars could be colossal for humankind.

Let us focus for a moment on cloaking, shielding and quarantines.
To protect the network of the future, we must anticipate malicious attacks well in advance. These attacks could be launched against the core infrastructure of the network, edge devices on the network, content/services, or all. A virtual quarantine isolates infected elements from the rest of the network, stopping the spread of the infection to other areas.

In the next release of this book, we will elaborate on the"what" for the "Predictive Cybersecurity" to accomplish this vision. Here is the high-level view...with access to large volumes of IP data traffic, we can carry out forensic analysis of network traffic using deep packet inspection (DPI). This analysis detects patterns that may be early indicators of a new worm, virus, or malware attack. This information is then used to isolate network elements from attacks in a proactive manner, and to isolate and quarantine specific infected elements. The key to the success of this approach is to have a direct way to communicate with the network elements and edge devices that are already infected. See Volume II of this book for the details.

4.6.4 FMO ETE Connectivity Architecture

Deploying and operating a DSP network is challenging. We need a network that behaves as one fully integrated and functional service delivery network, given the fact that the content maybe residing on other interconnected networks. Connectivity architecture is to provide interoperable communications among endpoints to enable component integration for delivery of end-user services at the expected QoE performance level.

The ETE connectivity architecture is comprised of the DSPs' numerous transport components, including the core, aggregation, backhaul, access, and transport interconnection with the sources of contents and other networks; it is also comprised of numerous content sources and processing complexes which are critical for enabling the reachability of the data sources; network security, user registration, and authentication of the users and the variety of the network components; content sources; security of the content sources; etc.

Based on our BII experience, the ETE connectivity architecture must address the following key aspects of connectivity: (a) Network Connectivity architecture (with the focus on optimum connectivity with the endpoints), (b) Service connectivity architecture (with the focus on QoE), (c) QoE probe architecture (with the focus on making the network performance visible with drill-down capability to the root causes), and (d) Standard node architecture (with the focus to recover from a disaster within a prescribed time frame for service restoration in case of catastrophic network events). In this section, we'll present a generic DSP ETE connectivity architecture.

4.6.4.1 FMO Network Connectivity Architecture

In the picture below (Figure 4.97), we are outlining the overall network connectivity architecture for the DSPs. This architecture takes into account the overall network from the customer premise, through the access network, through the edge and core for the on-net users, and the data centers for the on-net contents, and through the network interconnection to the off-net content and off-net users.

The DSP network is comprised of IP and optical transport endpoints. It is critical to have the control plane of these endpoints with an open architecture and north-bound-interfaces, each enabled by SDN and SDN controllers; also, a unified orchestrator is needed in order to coordinate both IP and optical SDN controllers and ensure service needs are being cared for in real-time. The end-points in this architecture are: users (wireless, wireline), access nodes, IP aggregation nodes and network, ethernet transport point-to-point and nodes, IP core, iCDN data Centers, POP data centers, services data centers, and gateway blocks.

The unified DSP network architecture must support the Quad-paly services (Wireless, wireline, video, and data). This network is comprised of a national core network (uCore), also, in some geographical areas, there may be justification to establish and connect to a regional network. Where a regional core is deployed, inter-city transport is used to create regional express links avoiding the national uCore. In this case, the design of the regional core must allow > 2 points of connectivity between the regional core and the uCore. With a regional core, DSPs can aggregate traffic and keep it local, in the region, to serve customers, and can establish scalability and resiliency given the regional and national level infrastructure. The complexity of the control plane of the uCore is ever-increasing (e.g., IP, OSPF, BGP, Multicast, RSVP, P2MP, etc.), care must be taken to segment the network and reduce control plane signaling, including BGP+IGP mechanism, label aggregation PLUS, multi-topologies in OSPF, etc.

The uCore transport network is a dedicated DWDM partial mesh network. Access to uCore is through dual-homing from a nationwide OTN partial mesh network. The OTN network is also used extensively for point-to-point layer 2 services. Given the ever-increasing traffic volume (50% increase YOY), triggered primarily by the video content, a significant speed upgrade throughout the transport network is required. This includes backbone and aggregation routes upgradable to 100G+, regional uplinks to 40G+. It is important to ensure that the aging DSP national and inter-city fiber network is manageable from impairment perspective, including PMD, Loss,

3B1 degradation, etc., to ensure that the transport network is upgradable to higher Gbps.

uCore is connected to the other international ISPs as well as other cooperating networks through GateWay blocks directly connected to uCore.

Aggregation network will be supporting quad-play services, dominated by interactive video and variety of like services. uSAG node's functionality is best to focus on (1) Traffic Aggregation, (2) Grooming, (3) Label Processing, and (4) Packet Processing. In this architecture, we have assumed a 2-level architecture for uSAG in order to enable the most efficient and economical aggregation of low-speed access ports to higher speed ports in uCore. uSAG network is a layer 2/3 IP network and backhaul to the layer 3 IP network through dual-homing to national uCore, and in some cases to the Regional uCore.

uSAG scope is from the uplink from the uAccess node to the uplink to the PE nodes. It is highly recommended that the MPLS boundary be extended, beyond the PE, to the uSAG layer.

The access network supports quad-play services, dominated by interactive video and a variety of like services over wired and wireless media. Access node's functionality is best to focus on (1) Aggregation function, (2) Grooming, (3) Backhaul to Aggregation IP Nodes, and (4) Backhaul to aggregation PL nodes.

uAccess (IAN platform) connects subscribers (wireline and wireless) to the immediate POP in the CSPs' network, and its' scope includes Customer Premise Equipment (CPE) and SIAD for wireless cell sites, CSPs' NTE/ONU collocated at the customer location, and the Integrated Access Node (IAN) in the COs' POPs. To preserve the legacy access, it is always the case that the existing FTTN, IPDSLAM, EoCU nodes are IAN upgradable.

The IAN performs edge packetization (for PSTN, TDM grooming, and TDM-to-IP conversion function, and pass-through for IP NTE, and cell site backhaul). It hosts various access technologies and can be expected to be deployed with the kind of technology that makes the most sense in the location it is placed. For example, IAN in the CO may have xDSL and PON. IAN at the SAI may have VDSL2 and optical Ethernet. The benefit of IAN is that it can reduce the disparate chassis, power, and trunking from the PMO as well as allowing homogenization of management, control, OAM, and trunking across services. This provides a layer of pre-aggregation as well as an interconnect among services that was not previously available.

This architecture is aimed at providing Ethernet over OTN for multi-service edge access, ethernet over OTN for a 2-level aggregation over OTN and delivery to the IP network, IP-DWDM over OTN for regional PE connectivity, and IPoDWDM for regional and national core networks.

In this architecture, the services residing in Unified Service Data Centers are made available through direct connectivity to the PEs; and in some cases,through the aggregation nodes, for the on-net customers; and through the peering PEs for the off-net contents. The logic for the placement of the contents (behind local aggregation nodes, local PE, Remote PE, and IXP node) must be based on the proximity to on-net/off-net eyeballs. Our best practice is indicating proximity to the on-net at > 40% traffic mix. This way, the content could be placed at local aggregation nodes, or local PE, or remote PE. For the off-net (short distance), the placement is most economical

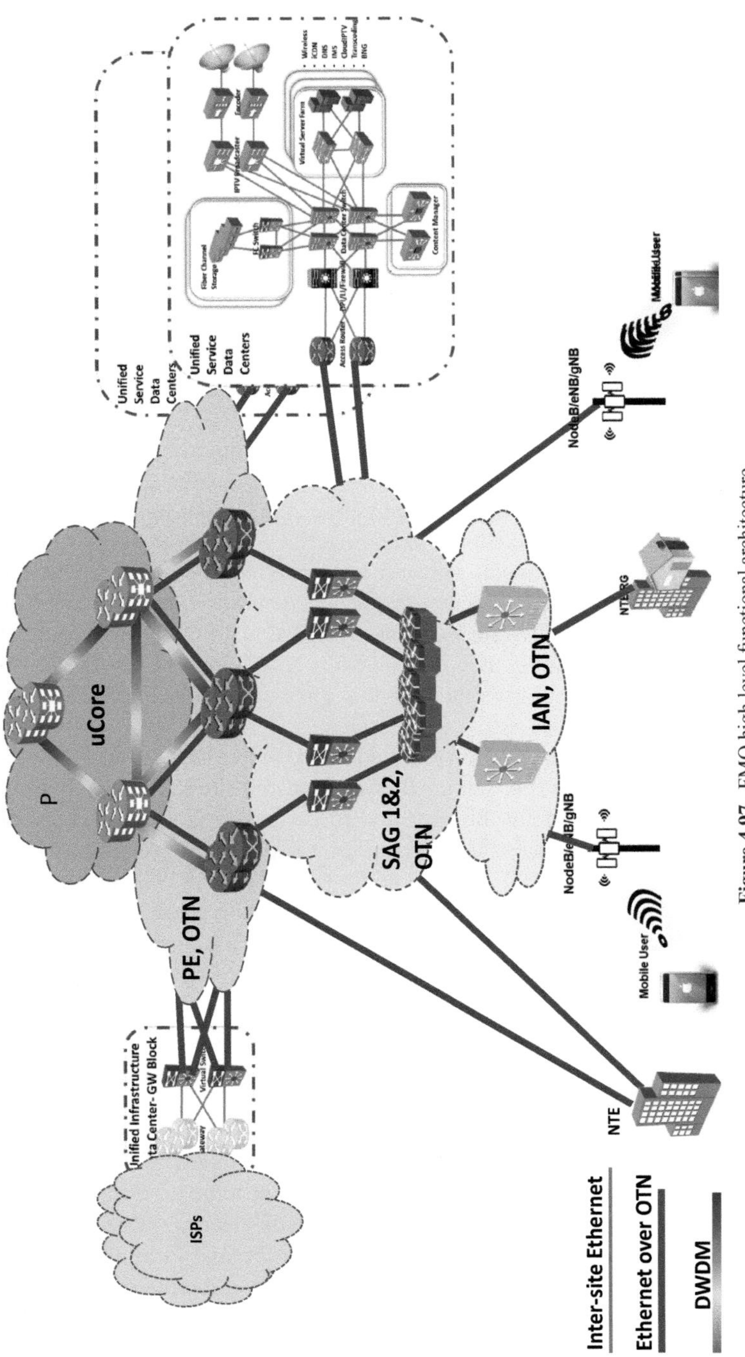

Figure 4.97. FMO high level functional architecture.

at the remote PE at > 60%, and for the off-net (long-distance over > 5000 kilometers) the placement is best at IXP PE at > 90%.

The unified Service Data Centers (i.e., iCDN and service complexes) are connected to three distinct nodes in the FMO: (1) uCore PEs in the DSP network, (2) Gateway Blocks connected to the IXPs and the International ISPs, and (3) uSAG's node in the aggregation network (Figure 4.97).

4.6.4.2 FMO Service Connectivity Architecture

DSP ETE services include content subscription services (e.g., IPTV, VOD, interactive video, news/magazines, music), connectivity subscription services (e.g., Internet access, VPN, E-line, E-LAN(VPLS)), data center services (e.g., Cloud, managed, colocation), multicast services (e.g., E-learning), voice services, and the future incarnation of such services.

In the picture below, we are outlining the overall network connectivity architecture for the DSPs. This architecture takes into account the overall network from the customer premise, through the access network, through the edge and core for the on-net users, and the data centers for the on-net contents, and through the network interconnection to the off-net content and off-net users.

High-speed Internet Access: Here is the service connectivity architecture for the High-speed Internet Access (Figure 4.98). A typical transaction flow for online banking or shopping will start with a search engine for the target web site, followed by name resolution, leveraging the DNS infrastructure, it then follows with the establishment of a secure connection (e.g., SSL/TLS) to the Certificate Authority, then the process for authentication of the user, and enables the banking/shopping transaction, and finally ends with the termination of the session.

On a national network of 5000 kilometers, and depending on the source of the content, being placed off-net, or on-net at a remote connected PE, or on-net at a regional connected uSAG/PE, the overall performance for latency is well below 85 ms, jitter well under 7 ms, and packet loss below 0.005% for access through Wireless LTE, xPON, TDM, SDH. At this level of performance, the most critical application for interactive video (e.g., gaming) can be supported with the expected QoE. It is important to note that the xDSL access is not recommended as an access technology going forward due to poor performance and limited bandwidth.

For video subscription services, there will be a wait time of single-digit seconds for buffering for the first chunk of the video, followed by a seamless transition to the subsequent chunks.

Teleworker/B2B Access: Here is the service connectivity architecture for a teleworker/B2B (Figure 4.99). Depending on the source of the business content, a typical transaction flow for a teleworker will start with a search engine for the target web site, followed by name resolution, leveraging the DNS infrastructure, it then follows with the establishment of a secure connection (e.g., SSL/TLS) to the certificate authority, this is followed by the process for authentication of the user, enabling the teleworker/B2B transaction, and finally ending with the termination of the session.

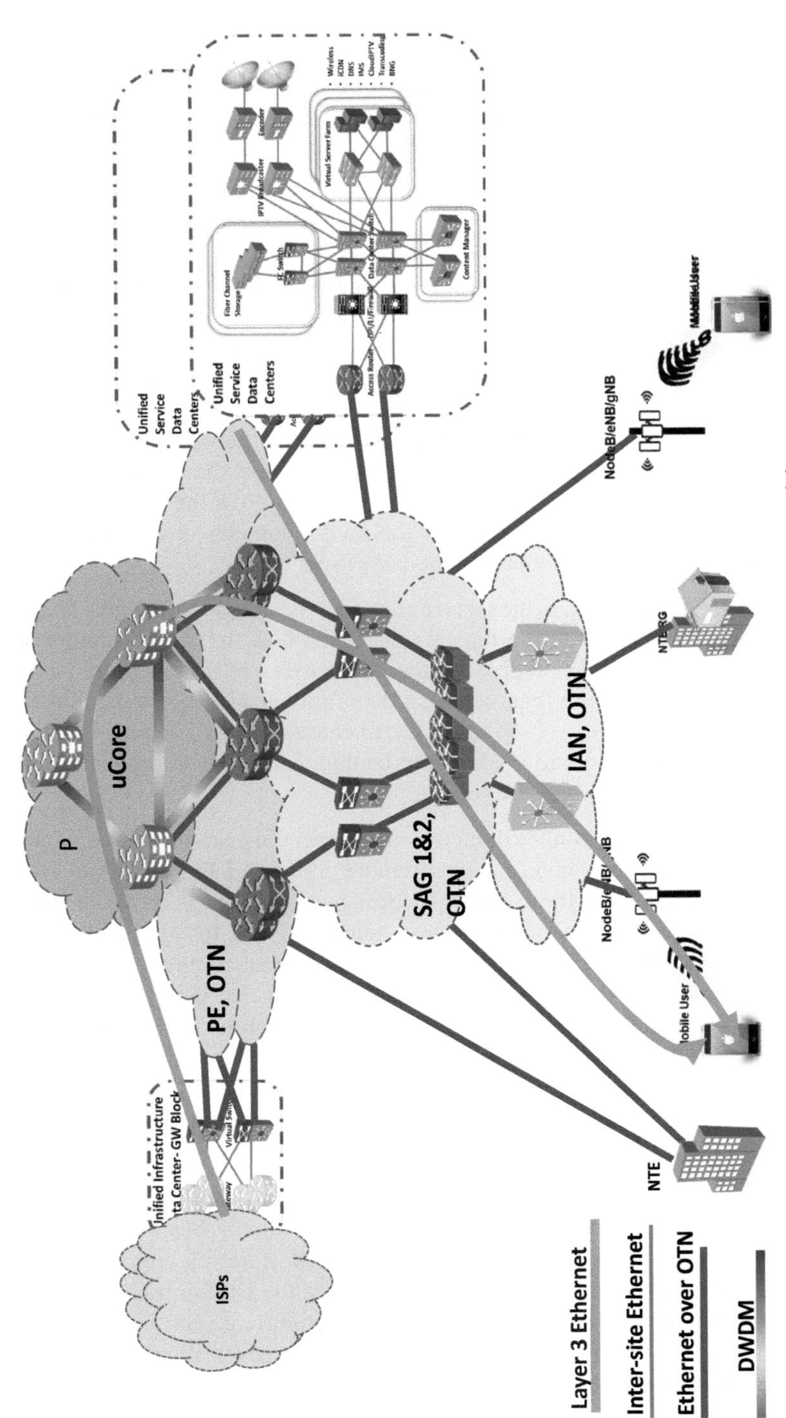

Figure 4.98. FMO high speed internet access connectivity.

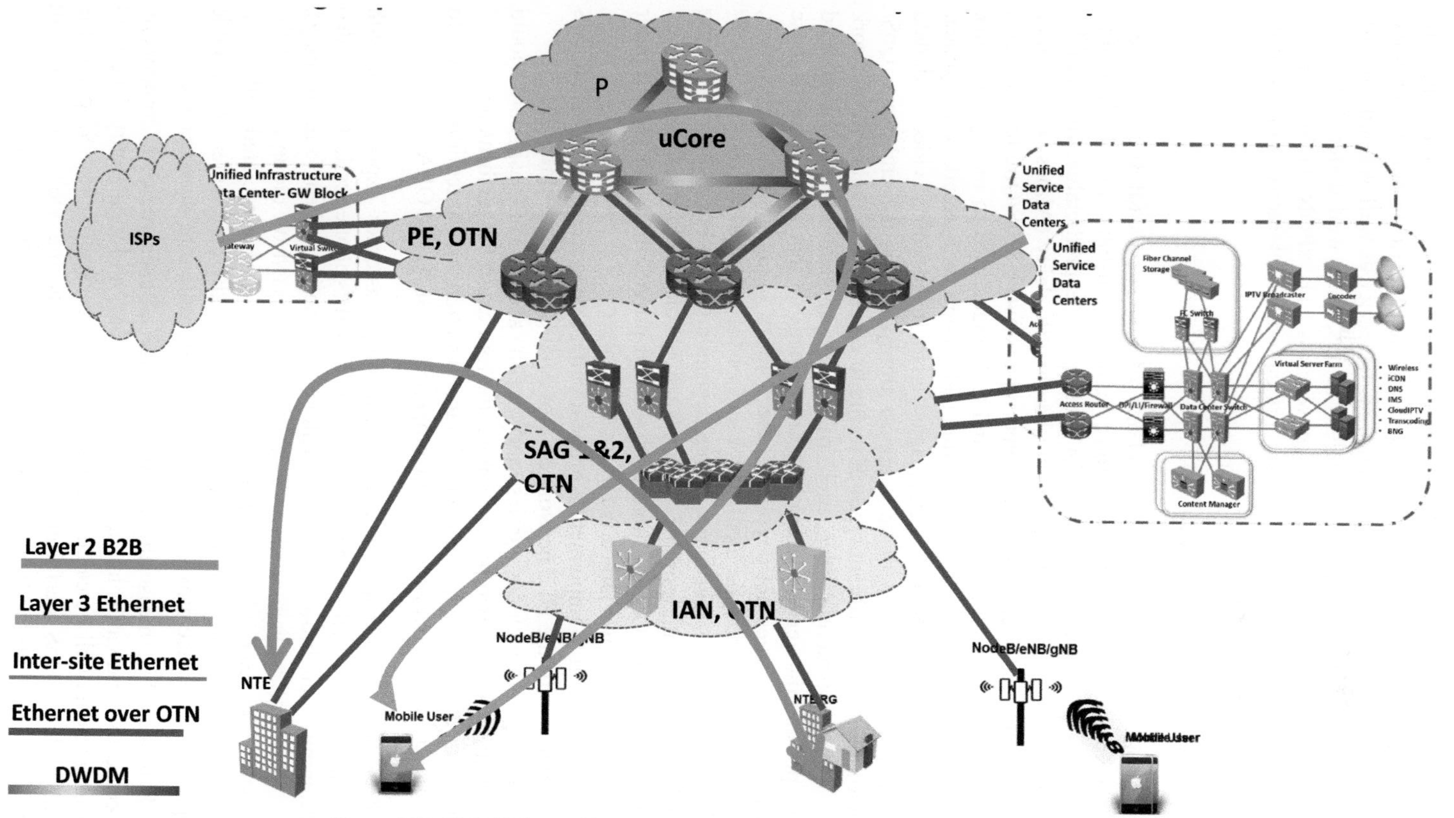

Figure 4.99. FMO high speed internet access —telework layer 3, BTB layer 2.

On a national network of 5000 kilometers, and depending on the source of the content, being placed on-net at a remote connected PE, or on-net at a regional connected uSAG/PE, the overall performance for IP service as well as point-to-point services is delivered with latency well below 85 ms, jitter well under 7 ms, and packet loss below 0.005% for access through wireless LTE, xPON, TDM, SDH. At this level of performance, the most critical application for interactive video can be supported by the expected QoE.

4.6.4.3 FMO QoE Probe Architecture

Communication networks form the backbone of any successful society. These networks transport a multitude of applications, including real-time voice, high-quality video and delay-sensitive data. Networks must provide predictable, measurable, and often guaranteed services by managing bandwidth, delay (Almes, 1998), jitter (Demichelis, 1998) and packet loss parameters on a network. QoS/QoE technologies (Bouraqia, 2019) refer to the set of tools and techniques used to manage network resources and are considered the key enabling technology for network convergence.

The objective of QoS/QoE technologies is to make voice, video and data convergence a reality, and to make it appear transparent to the end-users. QoS/QoE technologies allow different types of traffic to contend inequitably for network resources. Voice, interactive video, and critical data applications may be granted priority or preferential services from network devices so that the quality of these strategic applications does not degrade to the point of being unusable. Therefore, QoS/QoE is a critical, intrinsic element for successful network convergence. QoS/QoE tools are not only useful in protecting desirable traffic, but also in providing differential services to undesirable traffic, such as the exponential propagation of worms. We must use QoS to monitor flows and provide first and second-order reactions to abnormal flows indicative of such attacks.

It is highly critical to implement a probe architecture to measure QoE and QoS at every component of the ETE network and for each CoS in real-time.

The ETE network includes wireless and wireline access. This dashboard should also provide drill-down capability to pinpoint the root cause for performance.

Within the context of this aspect of transformation, our intention is first to describe QoE, QoS, "Network Provider Components", SLA, and OLA:

(a) QoE is aimed at establishing and measuring customer expectations with "connection-oriented" services, such as a VoIP phone call, OTT video streaming, web browsing, file download, interactive video (e.g., gaming), etc. The QoE measurements will be calculated based on the average of the 95 percentile peak values. What is excluded is the QoE associated with "non-connection oriented" services such as "a call to the call centers for trouble reporting", "service provisioning", or "trouble resolution" services, etc.

(b) QoS is aimed at measuring the performance of each network component, specific to bandwidth, latency, jitter, packet loss, and availability. The QoS measurements will be calculated based on the average of the > 95 percentile peak values. QoS performance directly impacts the QoE, i.e., we will specifically define the ETE QoS performance targets for each component of the network that must be met in order to deliver the right QoE for each key service. The network components for the wireless/wireline services include:

1. Access: xG wireless, Digital Subscriber Link (DSL) or Ethernet/IP, xPON, TDM, CWDM
2. Aggregation:
3. Core: PE-to-PE, PE-to-IXP
4. Content data centers (on-net and off-net): PE-to-USDC, SAG-to-USDC
5. IXPs: off-net-content to IXP

(c) An SLA is an agreement between the end customer and the CSPs (Cisco, 2005). This is a legally binding formal "contract". SLA is aimed at measuring the performance of the ETE service provider as it relates to the services for the end-users. It measures the ETE network performance, specific to bandwidth, latency, jitter, packet loss, and availability.

Here is the QoE/QoS probe dashboard architecture that we have implemented across a variety of networks (Figure 4.100):

Thisprobe architecture is comprised of three layers: (1) Data Plane, (2) Performance polling, and (3) QoE reporting and access.

(1) Data Plane layer:
At this layer we will utilize existing network elements, such as content/applications, routers, gateways, CEs, switches, wireless (cell tower, mobile users), xPON, and DSLAM to introduce appropriate probes for data collection. The performance data collection layer collects performance data, e.g., PE-to-PE, PE-to-GW, PE-to-On-net content, GW-to-off-net content, SAG-to-PE, IAN-to-SAG, and Customer Performance Processing Units (CPPU) is employed at the customer locations to report on performance measurements of the last-mile. The data collection is focused on network latency, packet loss, jitter, bandwidth, and availability. Deployment of the CPPU is critical for managed services that could be offered to the enterprise customers. This capability will include CPPU-to-CPPU based performance reporting. Also, in the case of consumer applications such as Cellular, xPON or DSL, we must deploy client applications on the end devices or NTE for traffic generation and performance data reporting.

(2) Performance Data Collection:
At this layer, we will deploy appropriate applications and databases for ETE polling, processing, and storage of performance data. The performance polling is real-time; however, the reporting aspect must be based on 95 percentile performance.

(3) Reporting/PortalLayer:
At this layer, we must provide an appropriate application for the capture and storage of the QoE data for key services, such as VoIP, Browsing, OTT streaming services, and gaming. Also, we must provide appropriate interfaces to accept and store ETE SLA targets, as well as allocated intra-DSP SLA targets.

Given the above inputs, as well as the actual performance data from the Performance Polling layer, one must provide a real-time dashboard to present (a) the

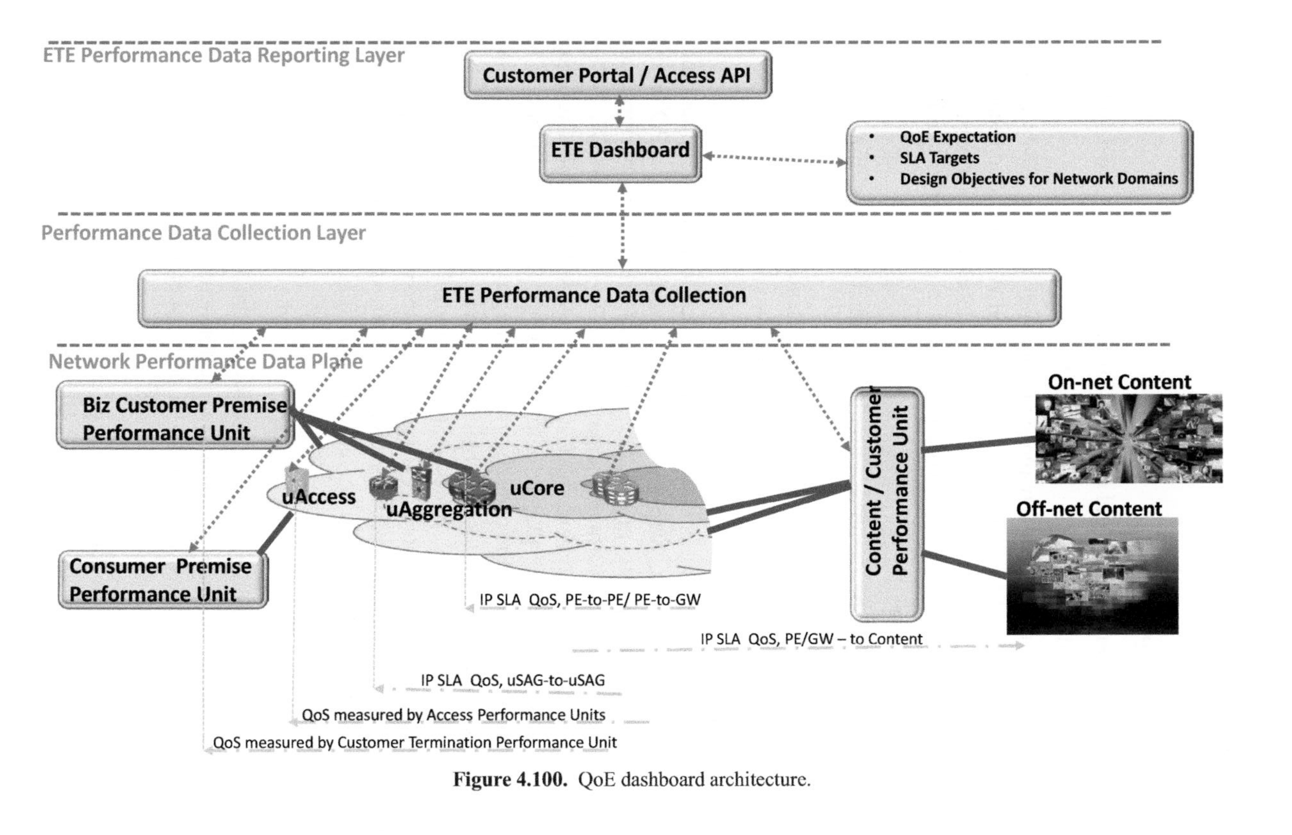

Figure 4.100. QoE dashboard architecture.

QoE expectations for each service, (b) allocated intra-DSP targets for each network provider components, (c) actual performance (QoE, and end customer SLA/Intra-DSP SLA) for each network provider component.

The probes are inclusive of the following types: HTTP ops, UDP echo and jitter ops, and ICMP echo, etc. This architecture must encompass the ETE network with a minimum of 15 network element types. As you see in the picture below (Figure 4.101), it is critical to incorporate a minimum of 3 initiator probes and 12 initiator/responder probes throughout the network, as depicted below.

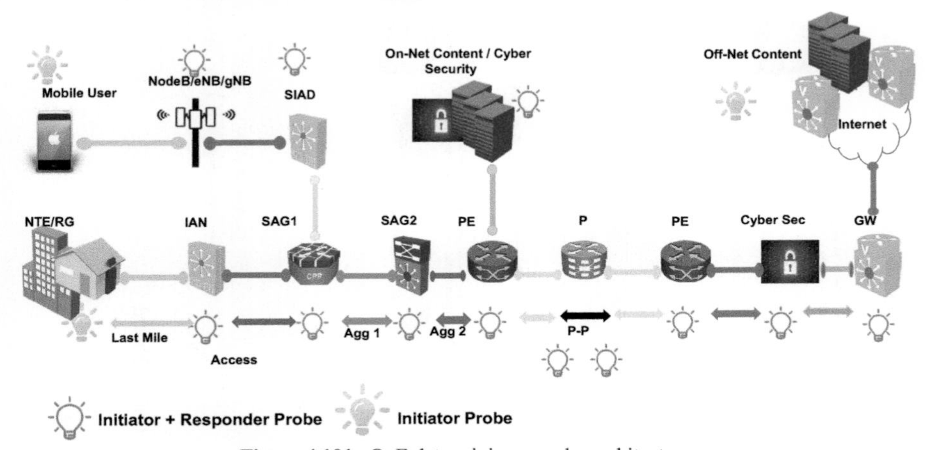

Figure 4.101. QoE data mining—probe architecture.

The probe architecture must incorporate QoS measurements (response time, latency, jitter, packet loss, bandwidth, and availability). The target network elements must include critical components of the ETE network, including the key elements given below (Figure 4.102):

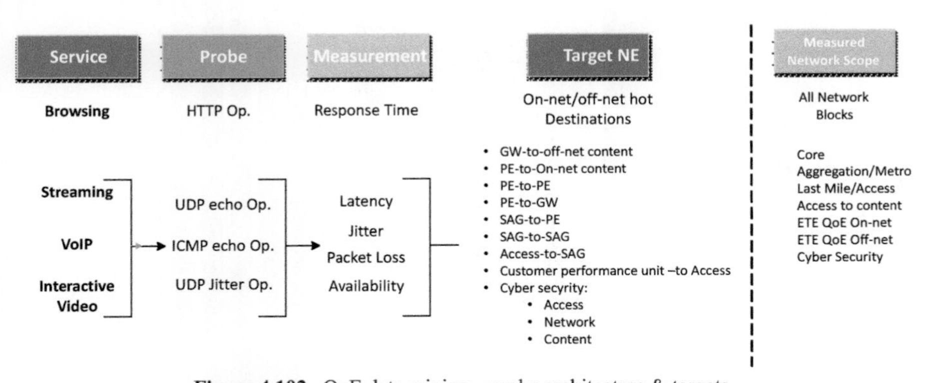

Figure 4.102. QoE data mining—probe architecture & targets.

The picture below (Figure 4.103) presents a typical QoE Dashboard. It is key to ensure this dashboard is cascaded in real-time to specific network elements and data content across the DSP footprint.

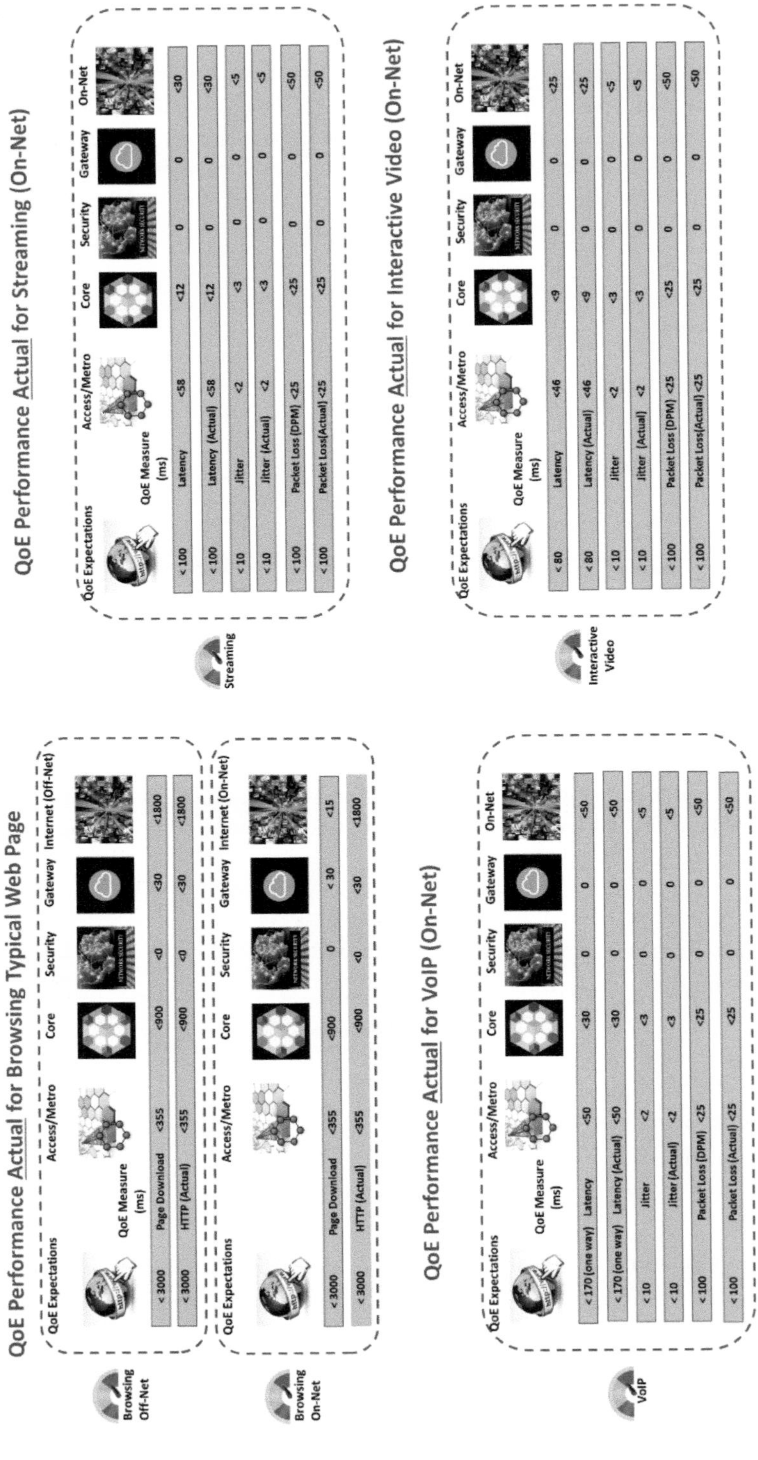

QoE Performance Actual for Browsing Typical Web Page

Browsing Off-Net

QoE Expectations	QoE Measure (ms)	Access/Metro	Core	Security	Gateway	Internet (Off-Net)
< 3000	Page Download	<355	<900	<0	<30	<1800
< 3000	HTTP (Actual)	<355	<900	<0	<30	<1800

Browsing On-Net

QoE Expectations	QoE Measure (ms)	Access/Metro	Core	Security	Gateway	Internet (On-Net)
< 3000	Page Download	<355	<900	0	< 30	<15
< 3000	HTTP (Actual)	<355	<900	<0	<30	<1800

QoE Performance Actual for VoIP (On-Net)

VoIP

QoE Expectations	QoE Measure (ms)	Access/Metro	Core	Security	Gateway	On-Net
< 170 (one way)	Latency	<50	<30	0	0	<50
< 170 (one way)	Latency (Actual)	<50	<30	0	0	<50
< 10	Jitter	<2	<3	0	0	<5
< 10	Jitter (Actual)	<2	<3	0	0	<5
< 100	Packet Loss (DPM)	<25	<25	0	0	<50
< 100	Packet Loss (Actual)	<25	<25	0	0	<50

QoE Performance Actual for Streaming (On-Net)

Streaming

QoE Expectations	QoE Measure (ms)	Access/Metro	Core	Security	Gateway	On-Net
< 100	Latency	<58	<12	0	0	<30
< 100	Latency (Actual)	<58	<12	0	0	<30
< 10	Jitter	<2	<3	0	0	<5
< 10	Jitter (Actual)	<2	<3	0	0	<5
< 100	Packet Loss (DPM)	<25	<25	0	0	<50
< 100	Packet Loss(Actual)	<25	<25	0	0	<50

QoE Performance Actual for Interactive Video (On-Net)

Interactive Video

QoE Expectations	QoE Measure (ms)	Access/Metro	Core	Security	Gateway	On-Net
< 80	Latency	<46	<9	0	0	<25
< 80	Latency (Actual)	<46	<9	0	0	<25
< 10	Jitter	<2	<3	0	0	<5
< 10	Jitter (Actual)	<2	<3	0	0	<5
< 100	Packet Loss (DPM)	<25	<25	0	0	<50
< 100	Packet Loss(Actual)	<25	<25	0	0	<50

Figure 4.103. Typical QoE dashboard.

The QoE dashboard must outline the QoE as follows for domestic/international services: browsing, VoIP, streaming, IV, bandwidth-consumer, bandwidth-business, service availability.

The service management graph for each network component is shown above. Samples are every 5 minutes and aggregated for predefined time intervals. The graph must show both 95 percentile value and average value. Each region/market must also show performance for the last mile access for wireless and wireline connectivity.

4.6.4.4 FMO Standard Node Architecture

Standard node architecture is a critical concept for (1) scaling the network, and (2) a speedy recovery from catastrophic failures. Please refer to the "Service Disaster Recovery" section for the details on standard node architecture.

4.6.5 FMO: The Processes/OSSs/BSSs

[This section is planned to be expanded in the next release of the book. The outline below is aimed to provide an introduction to what is planned to be included in the next release.]

Generally speaking, there are three categories of processes/OSSs/BSSs in a typical DSP: (1) Services over IP systems platform (SoIP), (2) Customer/partner-facing systems platform, and (3) Network-facing systems platform. A system platform in this context refers to the software/hardware/networking infrastructure upon which the platform is functioning. However, in this book, as we are not promoting a specific hardware,software or networking platform, we refer to a system platform by its functionality.

The overall architecture must be flexible and efficient in order to enable the fast introduction of new and emerging services (Figure 4.104). These services must be available to both business and consumer customers over any access technologies (e.g., wireless 4G+, WiFi, fiber, copper) while ensuring the same user experience, independent of the device or the access technologies being used. Also, it is critical to support the seamless handover of live sessions between any pair of supported access technologies or devices.

To accomplish this, DSP must transform its overall OSSs/BSSs architecture to a single, common, and shared infrastructure for all services and one service logic for all access technologies. This infrastructure must be based on standards, and must support a common run time execution environment.

Our intent in this book is not to describe the OSSs/BSSs architecture for a DSP. The key is to describe the overall functionality that must be put in place to accomplish the above architectural requirements.

Overall, the goal of the SoIP/OSSs/BSSs must be:

a. Fast introduction of new IP services through an open service creation system
b. > 98% Service Delivery ETE Flow through (Order through billing activation) in real-time
c. > 99.99% Billing Accuracy
d. > 99% E-enabled Customer Adoption Rate

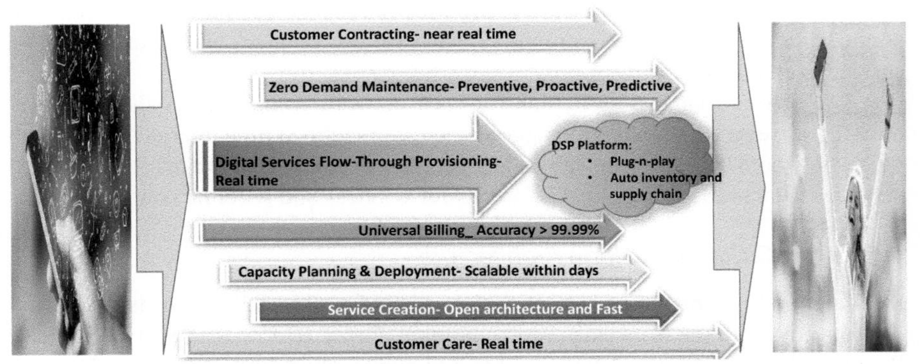

Figure 4.104. Concept of one/zero/none: OSSs/BSSs/Service creation transformation.

It all starts with the processes that must be designed and engineered to deliver on this performance level. There must be an all-out attack on the defects and cycle time when reengineering the processes. Extreme automation is what is needed, under the Concepts of Zero and None, to deliver such a tall order. It is doable, but must be orchestrated step by step in order to be achieved.

4.6.5.1 FMO DSP Processes—Preview

Figure 4.105 presents a high level view of key end-to-end FMO processes.

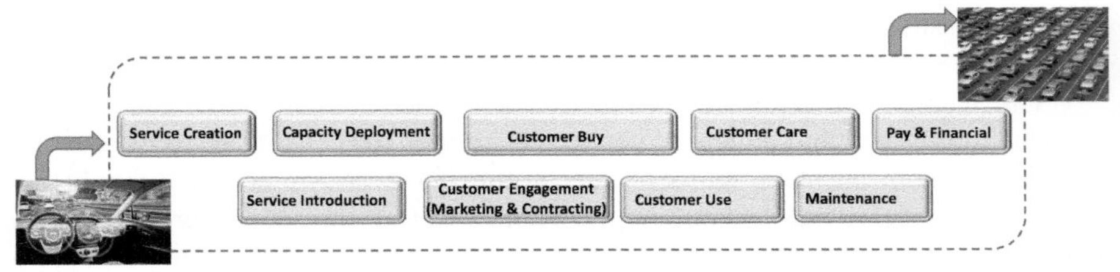

Figure 4.105. Conceptual DSP processes.

4.6.5.2 FMO Service Creation Processes—Preview

We have settled on two development processes to get ahead of the competition for the introduction of new services in response to customer demand for time-to-market services, (1) Agile, and (2) Water Fall processes. The project complexity must dictate what "development process" must be employed to get the job done.

The development process must be transformed to deliver in weeks/months vs years. There is no single development process that fits all types of development projects. The fundamental deciding factors for the use of these proven technologies must be found in the intent of what problem is to be solved.

The agile process is aimed at rapid development (< 3 months), with mostly feature enhancement (minimal to no data structure change, minimal to no database change, etc.). The requirements for these services are changing quickly and often, i.e., they often require small development teams, and are governed by a culture that thrives on chaos. This process works well with a collocated development team, and it is highlighted for development on top of the existing IT infrastructure.

Let us outline a typical example of what is not qualified for the agile development-when you need a new IT infrastructure (databases, data dictionary change, etc.) to get the job done, when the federated team is not collocated, when development requires more than 3 months to get the job done, etc. We used the agile process for all kinds of customer reporting, service bundling for telecom offerings, new services for differential seating offering in case of airlines, etc.

In a case where the data structure must be altered, the water fall process is the right process to use. In that, it starts with requirements, leading to design, and development, verification, user acceptance, and deployment. We need a single product management function to prioritize features, a centralized architectural decision making, with forward and backward traceability, and automated testing. The waterfall process is aimed at a new architecture, changing the IT infrastructure, e.g., databases, and data elements (with an extended development cycle time of >> 3 months where it is critical to deliver a 90% defect reduction that translates into 50% cycle time reduction). For these activities, the requirements do not change frequently, it often requires a large number of the development team members to accomplish the tasks, and the culture for these changes demands order.

Any process defects, if captured early in the waterfall process, say during design, require almost 10% resources for the rework, however, if they are caught during user acceptance, they require 90% resources to fix.

4.6.5.3 Service and Technology Introduction Process (Water Fall)—Preview

The highlight of the ETE Water-Fall process is shown in the picture below (Figure 4.106). This process is aimed at the introduction of new networking technology (layer 1 through 4+, new services, new IT infrastructure, staffing, training, leading to general availability throughout the DSPs' footprint).

The cycle time for this process is anywhere between 6 months to two years.

The BII practice for this process is to generate a T-Plan at the "Design" phase, followed by a Q-Plan to lock in all commitments, before the start of the "development" phase. The owner of the T-Plan is the technology lab of the DSP, and the owner of the Q-Plan is the program management for the Service Introduction.

Another BII practice in this process is the "Quality Gates". There are a total of eight quality gates in the process (Figure 4.107) to ensure the entrance criteria for the new step in the process are met.

4.6.5.4 Capacity Planning and Deployment Process—Preview

The highlight of the ETE Capacity Planning and Development Process is shown in the picture below (Figure 4.108). This process is aimed at the development of a Platform POR (Plan of Record) and an associated approved Capital Plan. This plan is to be built on an annual basis, and forms the basis for all capacity augmentation, footprint expansion, customer-driven demand, and new technology insertion transactions.

The cycle time for this process is to support (a) customer-facing transactions, i.e., for on-net customers at near real-time with point-and-click provisioning, and

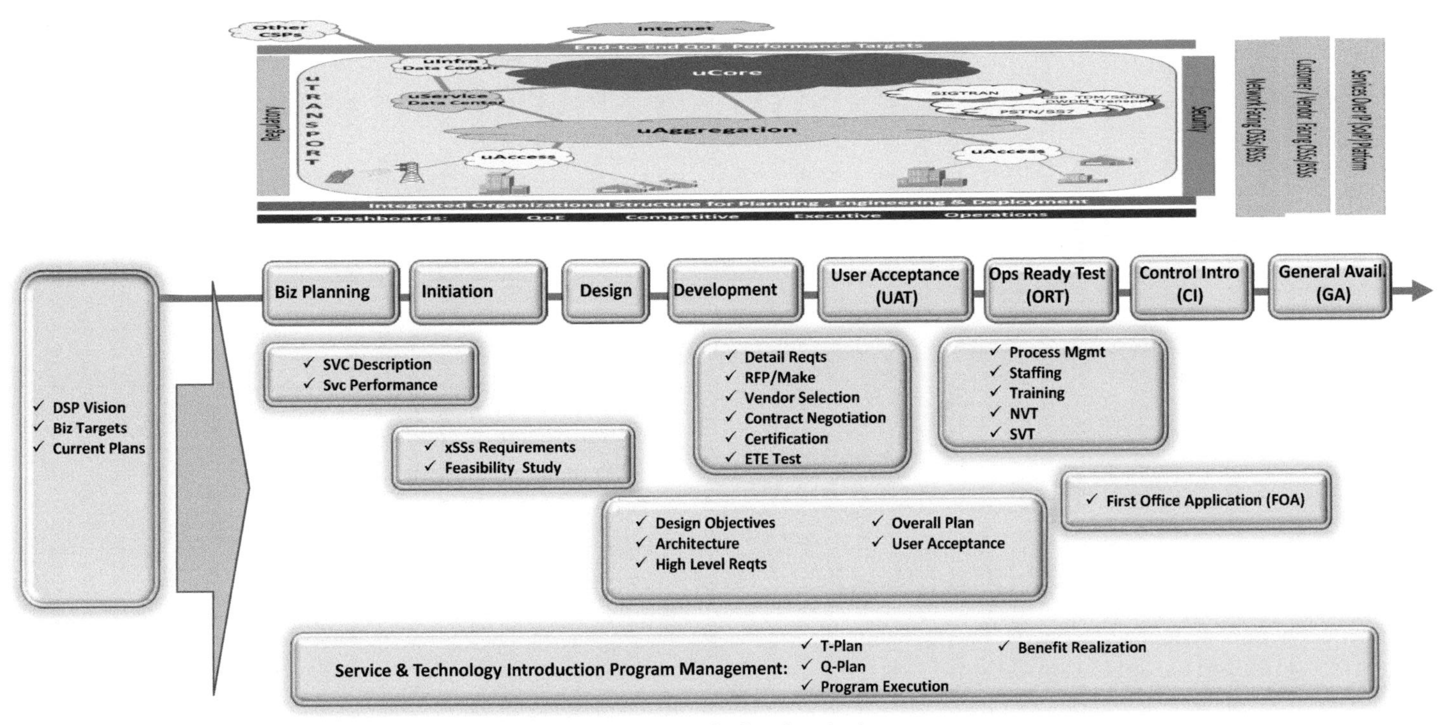

Figure 4.106. Service & technology introduction processes.

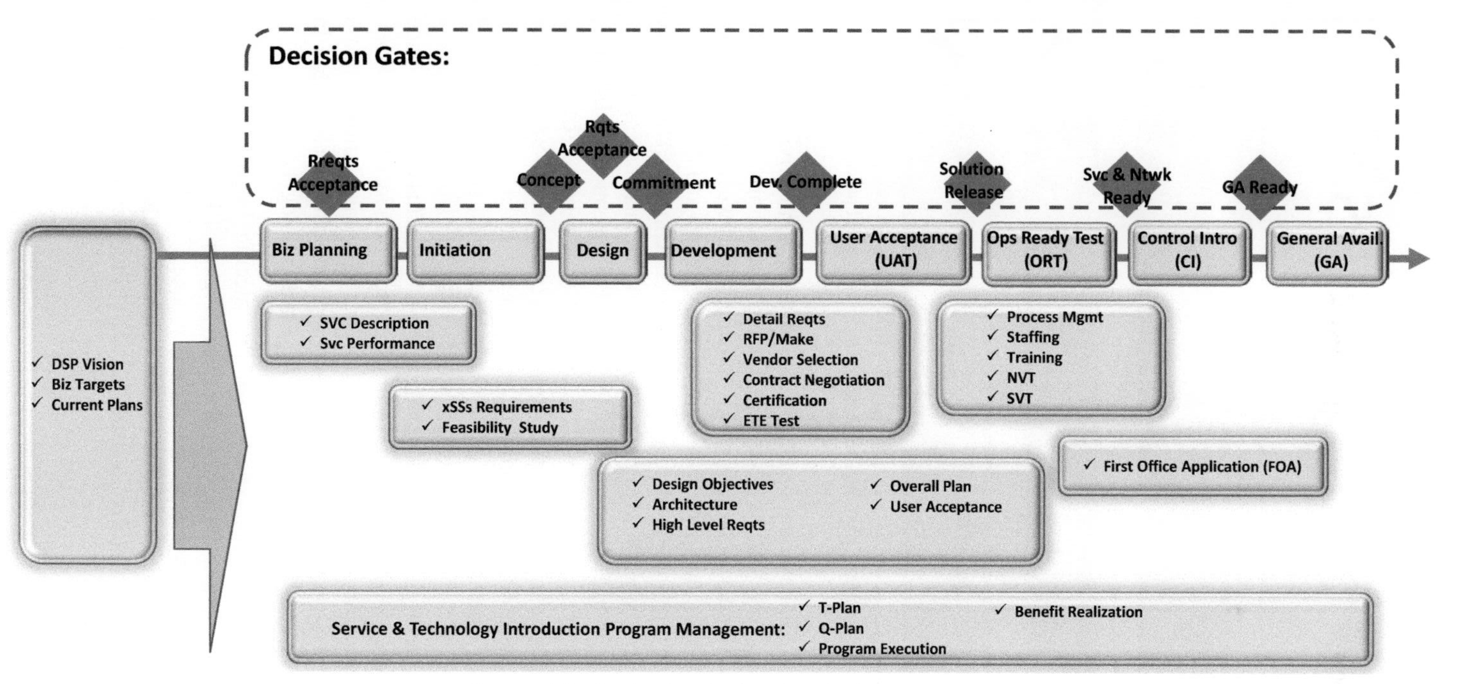

Figure 4.107. Service & technology introduction processes.

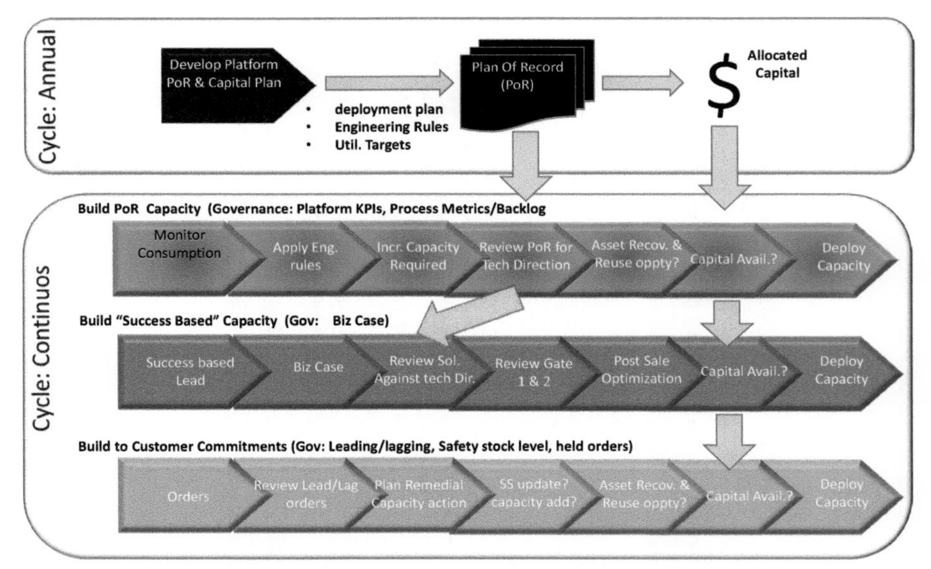

Figure 4.108. DSP network planning, engineering, and deployment process.

for off-net customers within days and weeks for establishing connectivity services, and real-time for content services; (b) Network facing transactions, i.e., network augmentation at same day interval, and any new construction at weeks or couple of months at most, including the acquisition of right-of-way (ROW) and the permits.

The BII practice for this process is to: (1) build and maintain the POR capacity per the specific engineering rules for capacity augmentation—in this process, it is key to ensure the capacity augmentation is consistent with the POR technology direction, as well as the availability of capital dollars (2) build the "success-based" capacity to specific customer built-to-order demand, and (3) build to customer commitment for GA services within the constraint of availability of the capital dollars.

The BII process for capacity planning and deployment must start with annual planning during the fall of the current year. In this process, the DSP must create a POR and capital program for the following year (Figure 4.109). This corporate-wide business planning must be based on the (a) volume forecast provided by the sales, (b) capital affordability provided by the CFO, and the technology/transformation direction provided by the labs and the transformation program management office. In the end, there must be an interlock on the POR and the capital budget.

(1) Build and maintain the POR capacity per the specific Engineering Rules for capacity augmentation throughout the year (Figure 4.110). All capacity deployment must be triggered to avoid capacity exhaust. There must be a tight management of equipment inventory and restocking intervals. It is also critical to assume a safety stock buffer in order to accommodate large orders and spikes in consumption.

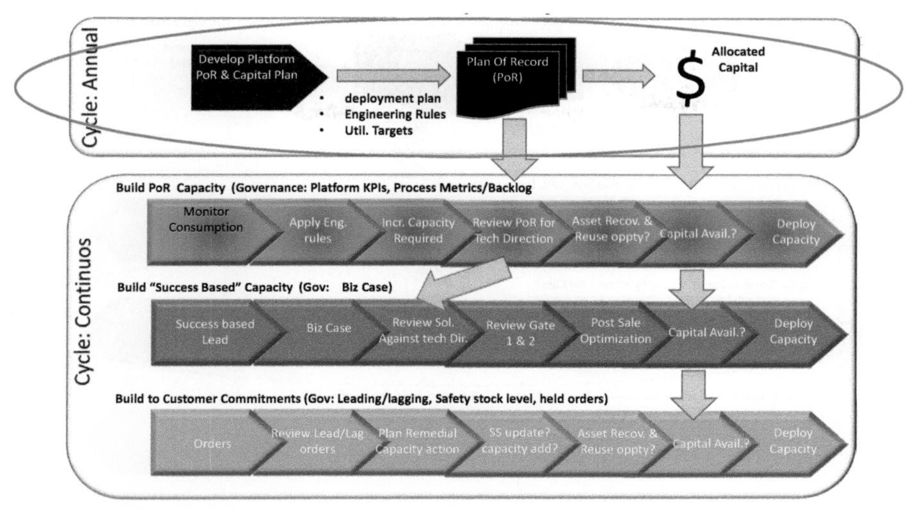

Figure 4.109. DSP network planning, engineering, and deployment process build POR capacity through the year.

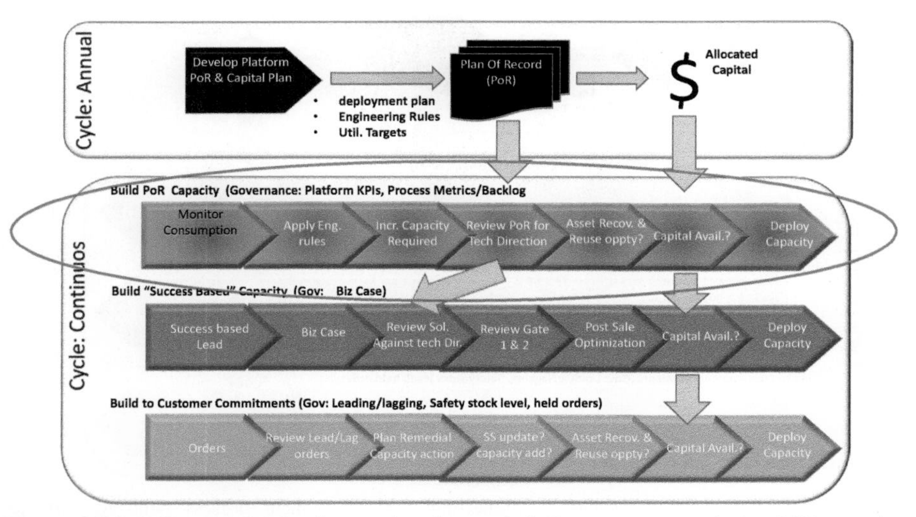

Figure 4.110. DSP network planning, engineering, and deployment process deploy POR capacity throughout the year.

(2) Build the "success-based" capacity to specific customer built-to-order demand (Figure 4.111). It is key to have a process in place to evaluate large customer opportunities, i.e., it is required to have a business case justification for the custom build to ensure adherence to the hurdle rates to justify the sale, and it is critical to have an architectural review to ensure availability of the technology.

(1) Build customer commitment for GA services (Figure 4.112). The objective of this process must be to ensure zero customer held orders. This process requires daily reviews of order status and weekly reviews of leading/lagging indicators to ensure that there is no missed commitment date. The leading indicators are

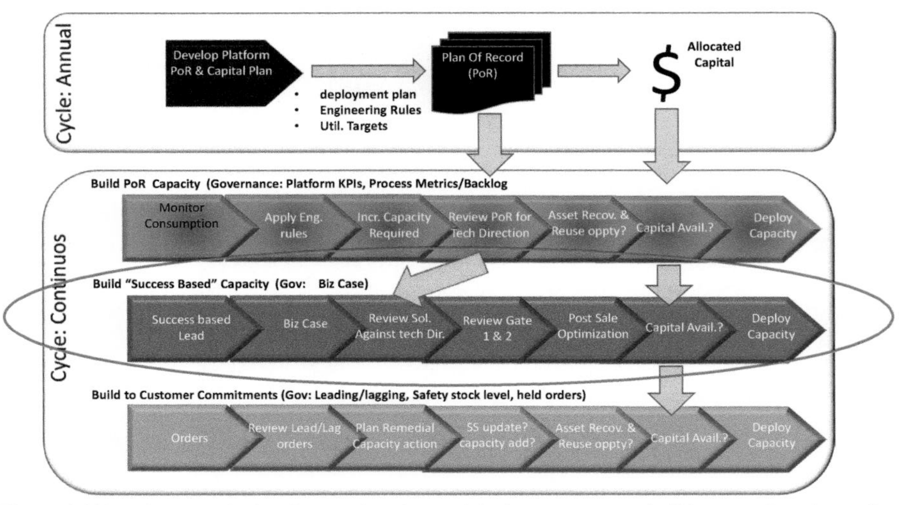

Figure 4.111. DSP network planning, engineering, and deployment process build success-based capacity.

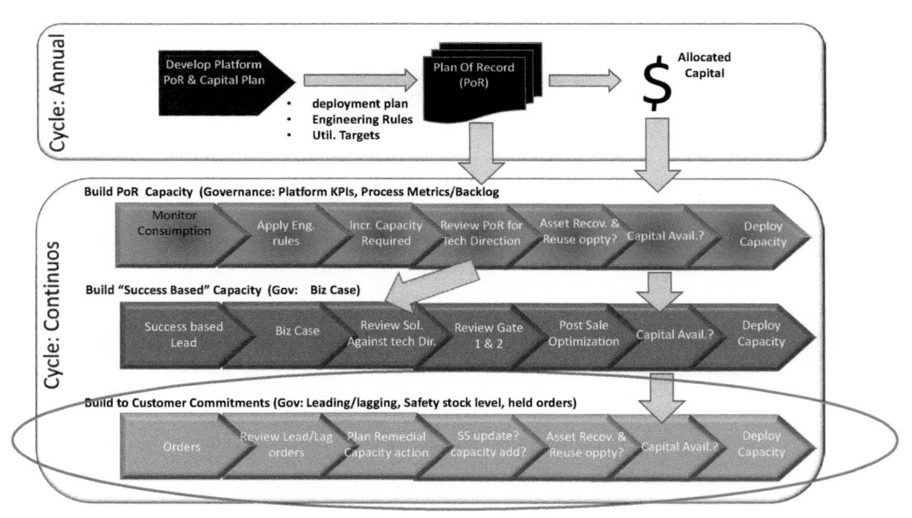

Figure 4.112. DSP network planning, engineering, and deployment process build to customer commitment.

aimed at highlighting the orders which are at risk for meeting future customer due dates. The lagging indicators are aimed at managing the orders that have passed the customer due dates.

The objective of this process must be to establish that there should be zero customer held orders. This process requires daily reviews of the order status, supported by the "leading" and "lagging" indicators.

When applying the AYS discipline to the capacity planning and deployment process, the following questions must be asked and verified before any capital expenditure made to supply the required capacity (Figure 4.113).

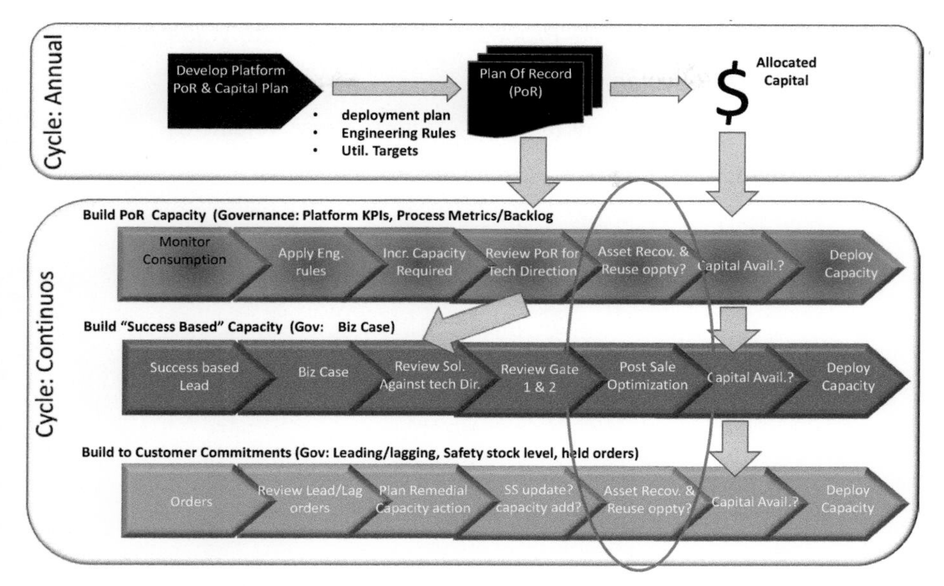

Figure 4.113. DSP network planning, engineering, and deployment process check (AYS capacity addition).

1. Have you verified that the demand is real (e.g., customer contract in hand, consumption trend verified, etc.)?
2. Have you challenged any reservations taking spare capacity?
3. Have all pending disconnects been worked?
4. Can you re-route or re-groom traffic to free up capacity?
5. Have you determined if the demand can be met economically and technically at a different location which has the capacity (e.g., backhaul, re-route)?
6. Have you determined if there is any existing physical asset that can be freed up (this includes warehouse spares) and moved economically?
7. What is the business impact if you don't get the additional capacity now?
8. Is leasing a better option?
9. If all of the above fail to satisfy the demand, what is the least expensive way to purchase the capacity with capital?

4.6.5.5 FMO Customer engagement: Marketing & Contracting Processes—(future release Volume II)

4.6.5.6 FMO Customer Buy Processes—(future release Volume II)

4.6.5.7 FMO Customer Use Processes—(future release Volume II)

4.6.5.8 FMO Customer Care process—(future release Volume II)

4.6.5.9 FMO Maintenance Process—(future release Volume II)

4.6.5.10 FMO Pay & Financial Processes—(future release Volume II)

4.6.5.11 FMO DSP Systems

Customer/partner-facing systems platform—(future release Volume II)

Network-facing systems platform—(future release Volume II)

CHAPTER 5

FMO Integrated & Real-time Dashboards

When we build the FMO closed-loop performance management system, the first and foremost driver is to make all DSP operations visible, and to be assured, as a result, that any potential shortcomings have a corresponding proactive mitigation plan in place and are executed, as planned.

Let's keep in mind that the scope of a tier 1 DSP is broad, and expands domestically and globally over many countries. Things do go wrong at any time, but no wrong should go for your customers.

As an executive, or an operation manager, or a process associate, the last thing you want to learn at 2:00 am on a Saturday, is that your global sales operations are being bottlenecked by the inability to make capacity available on time, or your customers are experiencing poor QoE performance in one or more operating markets, or a major business customer is experiencing a massive outage, or the cost of SLA compliance is exceeding your financials and negatively impacting the customer satisfaction, or your contra-revenue is eating into your free cash flow, and the list goes on.

The solution that we implemented time-after-time is integrated, real-time dashboards with the scope as indicated below (Figure 5.1):

Customer Facing Health	Network Facing Health	Economics	Effectiveness & Initiatives
QoE Performance	**Ntwk Performance** (e.g., access, aggregation, core, Data Centers, IXP)	**Capex unit cost**	**Competitive Performance**
SLA Performance	**Ntwk Security** (e.g., Core, Access/Aggregation, Content	**Opex unit cost**	**Vendor performance-SLA**
Biz Service performance (e.g., delivery, maintenance, assurance)	**ETE Process performance** (work volume, CT, Defects, Rework, Hand-offs)	**Cost of Cycle time**	**Network Improvement Initiatives**
Billing Performance	**Customer backlog performance**	**Cost of DPM**	
Customer care/sat performance		**SLA Compliance**	
Held order performance			

Figure 5.1. Dashboard—the scope.

The scope of the dashboard must cover the health of the customer-facing services and the network, the economics, and the effectiveness and initiatives.

Here is the scope of the dashboards that we are covering in this book (Figure 5.2):

Customer Facing Health	Network Facing Health	Economics	Effectiveness & Initiatives
QoE Performance	Ntwk Performance (e.g., access, aggregation, core, Data Centers, IXP)	Capex unit cost	Competitive Performance
SLA Performance	**Ntwk Security** (e.g., Core, Access/Aggregation, Content	Opex unit cost	**Vendor performance-SLA**
Biz Service performance (e.g., delivery, maintenance, assurance)	**ETE Process performance** (work volume, CT, Defects, Rework, Hand-offs)	**Cost of Cycle time**	**Network Improvement Initiatives**
Billing Performance	**Customer backlog performance**	**Cost of DPM**	
Customer care/sat performance		**SLA Compliance**	
Held order performance			

Figure 5.2. Dashboard—covered in the book.

5.1 Unified QoE Dashboard

We have already built the FMO network supporting the QoE expectations for the key 7 Macro Service Flows (Figure 5.3). Now, we are ready to build the QoE dashboard to monitor these QoE performance targets, as stated below (Halper, 2010).

Macro Service Flows	QoE Expectation	Acceptable Service Performance	Unacceptable
Browsing- Service flow	• Near real time page download	Page Download: < 3 Sec	Page Download: > 8 Sec
VoIP- Service flow	• Voice quality Same as wireline • Call set up cycle time same as wireline	Latency: < 170 ms Jitter: < 10 ms Packet Loss < 0.01%	Latency: > 250 ms Jitter: > 30 ms Packet Loss > 0.04%
Streaming- Service flow	• Zero pixilation • Zero freeze • Sync'd voice & video	Latency < 100 ms Jitter < 10 ms Packet Loss < 0.01%	Latency: > 120 ms Jitter: > 30 ms Packet Loss > 0.04%
Interactive Video- Service flow	• Same as streaming • Action Trigger sync'd with video	Latency < 80 Jitter < 10 ms Packet Loss < 0.01%	Latency: > 100 ms Jitter: > 30 ms Packet Loss > 0.04%
Bandwidth- Consumer	• Guaranteed throughput (DL/UL)	Guaranteed: Service flow bitrate /by market Symmetric speed on demand Bursting: 1G+ bps	No Guaranteed bandwidth
Bandwidth- Business	• Guaranteed throughput (DL/UL)	Guaranteed: symmetric, up to 40Gbps	< Biz demand
Service Availability (Biz.)	• SLA compliant • High availability	> 99.999 %	< 99.99 %

Figure 5.3. QoE performance benchmark during peak traffic hours.

The framework for the QoE dashboard must be based on a series of QoE probes (Figure 5.4): "Initiators", "Responders", and "Initiators/Responders". These probes should be established on the network elements and/or on dedicated probe devices throughout the ETE network from the customer premise to the contents, including the internet gateways, CDNs, and select off-net data centers. Below is the probe architecture.

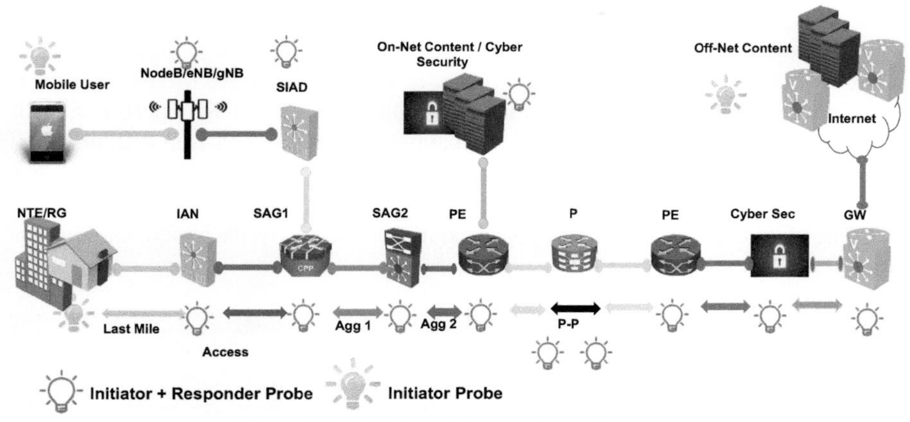

Figure 5.4. QoE data mining—probe architecture.

To measure the QoE performance, we employed minimum 4 key IP SLA probes (HTTP Ops, UDP echo, ICMP echo, and UDP jitter Ops) to measure response time, latency, jitter, packet loss, and network availability for QoE, as well as for every domain of the ETE network (Figure 5.5).

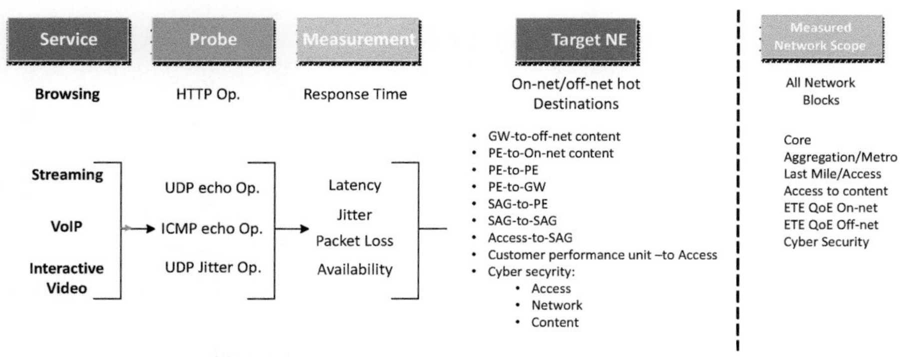

Figure 5.5. QoE data mining—probe architecture & targets.

The dashboard must summarize the ETE QoE based on 95 percentile performance. Samples are best collected every 5 minutes and aggregated for the last 12 samples in order to establish peak hours.

QoE must be expanded to the ETE network to provide coverage for the 7 service flows at the national level, down to the regions, in each operating market within the region, neighborhood, and specific customers.

The drill-down capability must be built to root-cause the poor performance to any specific market and specific domain in the network, e.g., access, aggregation, core, content services. Furthermore, it is essential to have the drill-down capability to the operations dashboard to see the specific actions taken to fix the problem.

5.2 Competitive Dashboard

The competitive dashboard is essential in ensuring that CSP is executing to maintain a competitive position in the market place. In this examination, we are looking to see if the competitor's performance is being monitored, if the CSP's performance is being captured and presented to enable an apple-to-apple comparison and, finally, if the business targets for the CSP are established and used to drive to a competitive position.

The picture below (Figure 5.6) is an example of what this dashboard should look like:

Market	Service	Service Type	Own Performance	BII Performance
All	**Internet**	Browsing (~9 MB Page download, peak hour)	15 seconds	3 seconds
		Streaming (peak hour)	Not measured pixilation Not measured freeze Not measured sync voice & video	Zero pixilation Zero freeze Sync'd voice & video
		VoIP (peak Hour)	Voice less than Wireline Unpredictable Call set up time Call Drop > .5% Texting greater than 10 seconds	Voice quality same as wireline Call set up time < 6 seconds Call Drop < .2% Texting within 10 seconds
		Interactive Video (peak hour)	Action trigger sync'd with video not measured	Action trigger sync'd with video
		Service Availability (peak hour)	99.99%	99.999%
Consumer	**Connectivity**	Speed (DL/UL): Wireless: Wireline:	 15 /4 Mbps 50 / 10 Mbps	 100+ / 20 100+ / 50
Business	**Connectivity**	Speed (DL/UL) CoS Jitter Latency	Symmetric 10Gbps None for ETE 30 ms ETE > 100 ms	Symmetric 10+ Gbps Multi tiers of CoS < 10 ms < 80 ms
	Strategic Content Services	Subscription VOD News Interactive Video	Minimal Some Minimal None	High High High High

Figure 5.6. DSP competitive dashboard.

This dashboard must be built to reflect on the QoE service flows, and must also measure the QoE during the peak hours.

Also, the service availability must take into account both the ETE network availability, as well as content platform availability.

5.3 Executive Dashboard (IBM Business Analytics for telcos, 2010)

The executive dashboard (IBM Business Analytics for telcos, 2010) is aimed at ensuring the calm that must exist over the operation of the ETE network. This dashboard ensures that:

- ✓ The preventive actions (such as a simple change of air filter on a router, or a software scan for the latest software version) are being taken at the right time to avoid network and service incidents.
- ✓ The proactive actions (such as potential capacity exhaust in a POP) are rule-based and executed on time, in anticipation of the customer demand, and,
- ✓ The predictive actions (such as the deployment of an optical amplifier to alleviate fiber degradation) are being pursued on time,avoiding long cycle corrective actions.

The executives must be engaged across the board, to monitor, and in case of potential systemic performance issues against the targets, are prepared to marshal company-wide resources to fix the problems proactively, before they turn into a systemic issue and a disaster.

Disaster is when DSPs have to gate the global sales operations due to lack of scalability and the supply-side issues, or to dance around an unpredictable sales uptick. Disaster is to face a global SLA compliance issue such as latency as it could impact global customers. Disaster is to face a systemic QoE issue in one or more operating regions. The DSPs' executives must take an active role in the performance of the overall network to avoid any disaster.

The closed-loop performance management system cycle (Figure 5.7) starts with (1) collection of alarms/measurements from the survivable network, (2) translation of the alarms into the dashboard, (3) surveillance, analytics, and notification for actions, and (4) restoration, repair, and verification actions, (5) capacity and technology actions, and (6) regulatory actions.

This dashboard will alert the executives to any deviation from the performance targets, through R/G/Y, and to enable the executives to ensure that network and services are being cared for in a proactive, preventive, and predictive way. The executive dashboard is best to include the content as depicted in Figure 5.8 and 5.9.

5.4 Operations Managers & Working Level Dashboard

The working-level dashboard is key in enabling the process associates for real-time decision making to care for the health of the network. This dashboard must incorporate the facts on the targets, actuals/results, alerting targets, and any appropriate actions that are taken and documented to proactively address the potential issues. This dashboard must be linked up to the QoE dashboard. The drill-down capability from the QoE dashboard to the working level dashboard is essential to ensure calm and control (Figure 5.11).

The operations managers and working level dashboard is best to include the content as depicted in Figure 5.10.

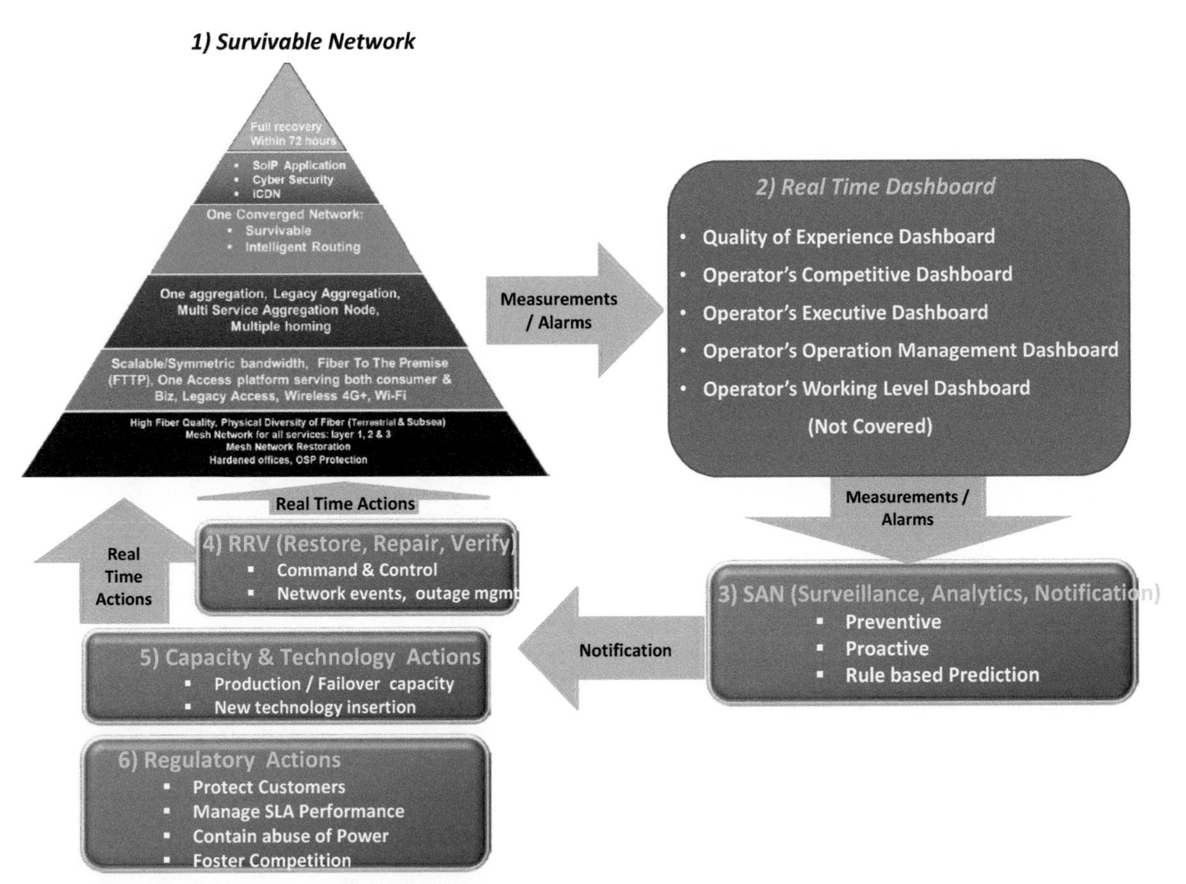

Figure 5.7. CSP's closed loop performance management system.

Exec Dashboard Content	Score Card
KPI Description	●
KPI Results	●
KPI Targets	●
Alerting Targets	●

Figure 5.8. Executive dashboard content.

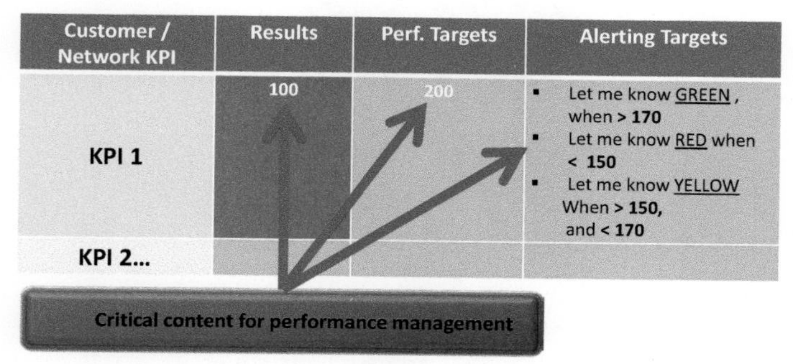

Figure 5.9. DSP executive dashboard.

Exec Dashboard Content	Score Card
KPI Description	●
KPI Results	●
KPI Targets	●
Alerting Targets	●
Action Target	●
Action Taken description and timing	●

Figure 5.10. Operation managers & working level dashboard content.

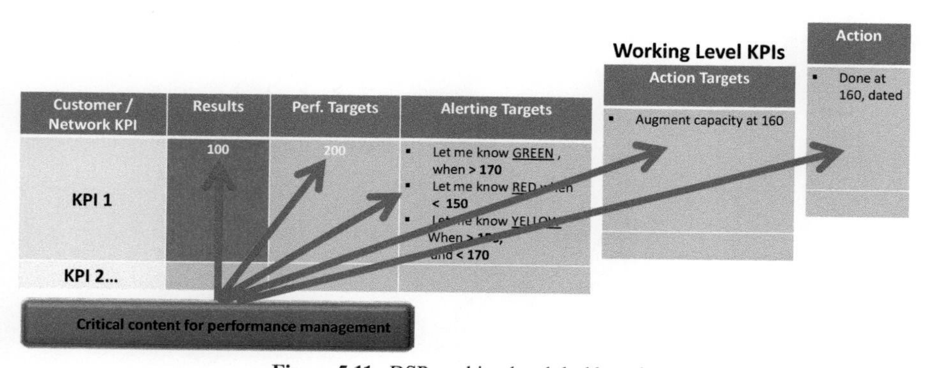

Figure 5.11. DSP working level dashboard.

The drill-down capability of the QoE dashboard to the DSPs' network elements (Figure 5.11.1) is critical in establishing the root cause for the service degradation down to a specific network element.

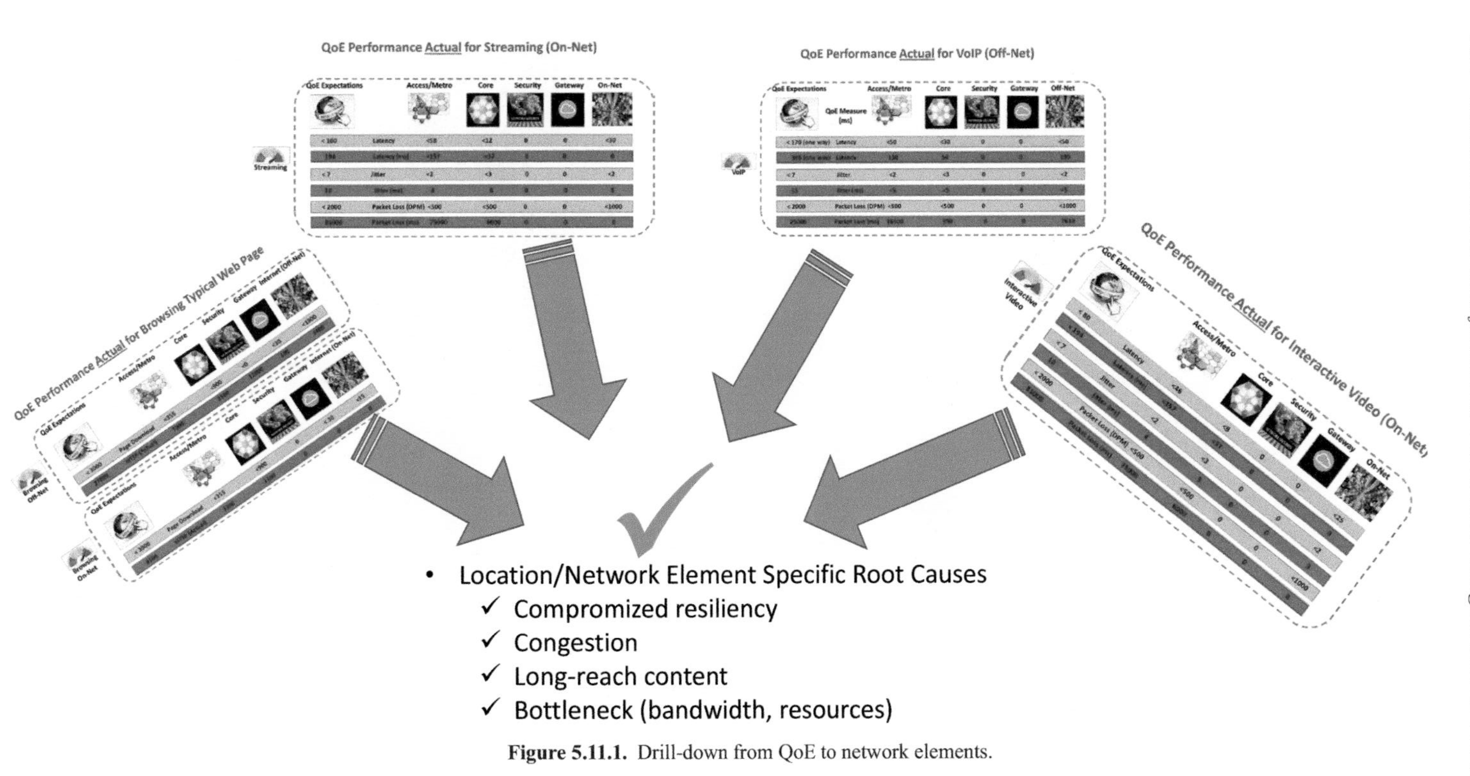

Figure 5.11.1. Drill-down from QoE to network elements.

CHAPTER 6

FMO Operations-Network & Services

Regardless of how perfectly your overall network and systems architecture are designed, regardless of the state-of-the-art networking and IT technologies that are deployed in your network, the overall QoE is at the mercy of how well your operations' associates are executing 100,000's of network impacting procedures (e.g., provisioning, maintenance, technology introduction, security defenses, network reconfiguration, etc.) on a day-to-day basis.

Under the FMO, you must drive your operations for *flawless execution*. It is all about following through with the BII operations practices, to deliver on the design objectives for the QoE and QoS,this includes reducing defect rates and cycle time through customer communication, network events, TCB/MCB, root access, outage management, methods and procedures, ask yourself, etc.

A typical operations process associate, under the FMO, must experience the following on a daily basis. I am a DSP process associate and...

(A) **Working Level Dashboard**—I am looking at the real-time network performance dashboard to establish what customer provisioning/configuration activities are to be executed, if the network is to be augmented, or fixed, or a routine maintenance is to be deployed, and if so, I follow the *ASK Yourself principle* to get the job done...in doing so, I use the *M&Ps*. My *authorizations to get the job done are all rule-based* and I do not need any other authorization before I get started. Also, *the root access to the entire network is exclusively owned by the operations*—no labs/R&D have access to the live network and no surprise whatsoever.

(B) **M&Ps**—There are times that a network event is scheduled, such as a software upgrade, and I make sure that *I have all the M&Ps available and approved to get it done, and I will follow the M&P to execute the steps as outlined.*

(C) **Ask Yourself**—Whenever, I touch the network, I first ask myself if I have gotten the training to do so. Also, I check for the schedule of authorization for the network touch to ensure that the timing to get this work done is authorized.

(D) **ASK Yourself**—I check with the NOC to make sure that my activities are being monitored when I am touching the live network. Also, I have checked and assured that the backout procedures are in place, in the event of unpredictable service interruption.

(E) In case of large-scale change, I do it under planned restoration with no service interruption.

(F) If there is a new service /technology introduction activity, I make sure that the NVT/ORT is completed and approved before the services are placed into a state of general availability.

This mode of operations requires an enablement, i.e., sophisticated set of BII processes, systems, closed-loop performance management system, and a workforce motivated and enabled to care for the end customers.

Chapter 7
FMO
The Regulatory

DSPs will continue to play a significant role in establishing a digital economy. DSPs are capital intensive, and their transformation from CSP to DSP requires significant innovation and investment, which is highly influenced by the uncertainties in ROIC. It is our collective view that DSPs must be regulated to unleash the necessary investment required for transformation to a DSP status. The challenge is in scalability of regulated directives which maybe possible through the application of AI. Regulatory agencies can and must help CSPs in their transformation efforts toward DSP. Regulatory agencies must provide leadership in developing standards to enable DSPs to execute content cleansing/censorship, without stifling the freedom of speech in the regions of the world where it is applicable.

Even though the trend for CSPs' platform convergence has been around for many years, the regulation has not converged and is continuing on the path for silo-specific regulation, with much more emphasis placed on regulated communication, and not as much on content and computing. This must change to enable liberization of the CSPs to become DSPs.

The main jobs of the regulatory institution are many-fold (ROSSI, Luisa, 2014), but it must be noted that it is critical to encourage innovation for the converged digital services. The key aspect of the regulatory engagement is to: (a) protect customers (short term and long term), (b) establish and enforce strict QoE/SLA management, (c) contain abuse of market power, and (d) foster competition and innovation.

Regulatory and competition authorities have a significant challenge ahead to enable the transformation. We believe that the regulatory authorities, as they practice transactions today, pay minimal attention to the effects of their decisions on the market concentration/vertical integration (à la DSPs), as well as innovation.

We believe the resolution of the following key issues, by the regulatory authorities, is essential to unleash the necessary investment toward a DSP status:

i. Liberalization of CSPs to become DSPs.
   One of the key drivers for a CSP's transformation into a DSP is the introduction of digital services (anytime, anywhere, any device) such as VOD, interactive video, news and magazines, etc., which in the age of technological convergence is stifling competition in the long term. For example,

A. Congestion of the CSPs' infrastructure as the margins, market power, and the capacity for investment is being reduced. Also, it is a key issue to encourage capacity adequacy of the public internet.

B. The current worldwide roaming and wholesale access pricing is anti-competitive and is not encouraging competition.

C. Consumer protection—It is proving extremely difficult to ensure the short-term price cap policies are not restricting the opportunity to undermine the digital future in the medium to longer term. For example, it is a huge investment challenge to prepare for broadband connectivity through 5G. Regulatory agencies must encourage densification of FTTx to enable ubiquitous coverage for the 5G+ signals.

D. It is challenging to stop dominant DSPs who could be stifling competition through acquisitions. It is proven that the dominant DSPs have the incentive for anti-competitive behaviour and abuse of dominance through acquisitions. For example, Google, through the acquisition of Beat-That-Quote, sought to promote its product to the detriment of its rivals, to exclude others from the market and maximize its profits from being one of the very few players left providing online insurance quotations. It is highly desirable, for the regulatory authorities, to monitor the acquisitions of all dominant companies for their effect on innovation.

E. Facilitating and promoting the production of original content. Also, it is important to address the production of original content relative to the streaming channels.

F. Unreasonable taxation issues associated with DSPs with global footprint and presence.

ii. Defensive Cybersecurity.
Cybersecurity is vital to a properly functioning and prosperous economy. Security vulnerabilities of the CSPs and DSPs are critical and are on the rise. These companies may choose not to invest adequately to stay ahead of the rising threats. Regulatory institutions must enforce the following mandates:

A. Regulatory initiated security assessments through penetration testing. The outcome and the associated learning from this approach will be most welcomed, and acted on by the private sectors, as well as the CSPs and DSPs.

B. Regulatory agencies conducted security assessment, for example:

(1) Typical network focused: Firewall, DDOS, two-factor authentication, ACL, profiling and data mining connected network device behavior, wireless intrusion detection system, wireless advance encryption, remote access through VPN, etc.

(2) Info system&content focused: limit USB connection, control access to external devices, encryption of data.

(3) User access management-focused: centralized authentication system, enforce expiration on all user accounts, force re-login process, 2-factor authentication, ACL, etc.

iii. Regulated DSP data Privacy.

A. Consumer protection in terms of privacy of personal information. Right to opt-in/out of data sharing/marketing, rights for the transfer of personal information.

B. Use of consumer personal information. Right of access to personal information, right of deletion/rectification of errors.

C. Regulated use of Cookies to ensure no attempt made for "deep packet inspection" of personal user data, such as passwords.

D. Bullying on social media.

iv. Regulated Internet Content.

Content purification is critical for liberalization of CSPs to become DSPs. The absence of such regulation has disastrous consequences as the unregulated DSPs are programmed and prioritized for monetization of end-user engagement, driven by advertising revenue from all possible sources: legal/ethical and illegal/non-ethical.

A. DSPs should be kept accountable for enforcing standards on harmful/violent content. As said before, regulatory agencies must provide leadership to develop standards to enable DSPs to execute content cleansing/censorship without stifling the freedom of speech in the regions of the world where it is applicable.

B. DSPs must have a proactive process in place to ensure election integrity through verification of political actors.

C. DSPs must be regulated for advertising.

D. Bullying on social media is not acceptable.

CHAPTER 8
FMO
The Transition Mode of Operations

8.1 The T-Plan for Bridging to the FMO

Transformation from a CSP to DSP is a multi-year, technology-transfer program. Through this program, the CSPs must build a series of capabilities/technologies (e.g., SDN capable network, unified IP/MPLS core/aggregation, VPLS enabled Edge, VPLS enabled aggregation, SoIP platform, NFV, OSSs supporting new network elements, workforce training, etc.) across all processes and the network platforms.

Once these capabilities are developed and verified, new services (e.g., VOD, subscription-based news/music, VPLS, Quote-to-cash SoIP application) can be built on top of them.

To manage this massive technology transfer program, we have developed the BII practice. It includes three major components: (a) a T-Plan (Time Plan) with a logical progression to build the technology capabilities, (b) a Q-Plan (Quarterly Plan) committed plan for network capabilities, and (c) a Q-Plan with commitments for service introduction.

The T-Plan is an architecture view for the network capabilities to be developed over the course of the transformation time window, up to a maximum of 3–5 years. It is important to note that the T-Plan does not represent the funding commitment. Also, this plan should be developed and owned by the DSPs' labs.

The Q-Plan is a time-based view for the network capabilities outlined under the T-Plan. It is important to note that the Q-Plan does represent the funding commitment, and it is committed by all DSP's processes impacted by the transformation, including the vendors/partners. This plan must be developed and owned by the program management office for the transformation.

The T-Plan is a collection of capabilities that must be used by the DSPs' services over time to fulfill the full scope of their target network and services offering. The grouping and phasing of such capabilities is based on: (a) availability of the technologies over a time window, (b) key business drivers for revenue generation and cost reduction, and (c) the sequencing required for technology insertion.

The T-Plan is composed of T-Steps. The T-Steps do not define the services that will be deployed, but the network capabilities required to build them.

In the context of the T-Plan, the T-Steps or the capabilities for the transformation could be outlined in terms of the steps leading to the development of key capabilities, such as QoE Dashboard, TCB/MCB, common transport layer, CDNs, data centers, uCore, uAggregation, uAccess, E911, workforce training, etc.

The essence of the T-Plan is to construct a step-by-step picture for the critical steps that must be taken to bridge the gap from the PMO to the FMO (Figure 8.1). We group these T-Plan steps into a collection, called "Transition Mode of Operation (TMOs)".

Here we outline the technology transfer through five TMOs to the FMO for a typical CSP that we led to transform. Please keep in mind that the number of TMOs is the function of individual cases for each CSP's transformation.

In this example, we are assuming PMO is primarily focused on connectivity services, who are suffering from poor QoE performance; its operations and services are not customer-centric (i.e., it is product centric); the CSPs are struggling with a high unit cost for CAPEX and OPEX, and their connectivity services are limited and have a thin margin.

Figure 8.1 presents a picture of a typical CSP before and after the transformation. Figure 8.2 outlines the major steps that are taken to transition from PMO to FMO through TMOs. In this transformation, poor QoE associated with PMO will improve dramatically, low speed connectivity services will be replaced by high-speed, and low margin connectivity services will be enhanced significantly by the inclusion of digital services. It is important to note that an extreme transformation such as this should not and will not have any negative impact on the end customers, throughout the journey through TMOs to the FMO.

Here are typical TMOs (Figure 8.2) at a glance for many CSPs for whom we have led the transformation:

8.2 Transition Mode of Operation 1

TMO1 is the most challenging step in the transformation. In this TMO 1 (Figure 8.3), we are establishing the foundation for the transformation with optimization of the legacy networks for the delivery of QoE, and fact-based visibility into the QoE for the senior management, customers, and the employees.

Before TMO1, we must have already completed the assessment and baselined the PMO with baseline QoE and KPI performance. We must have baselined performance dashboard for the PMO. This is the starting point and the reference point for progress to the FMO.

First, we must establish the targets for QoE for every legacy service that the CSP is offering. These targets must be cascaded to the KPI targets for each domain of the legacy networks for each TMO. Next, we must develop and deploy the QoE dashboard across all ETE legacy networks. It is a key factor for the success of the transformation to demonstrate that the CSP is transparent with the transformation, for example this dashboard must be made available to the customers, highlighting the performance gaps that the CSP is planning to close and indeed is closing over time.

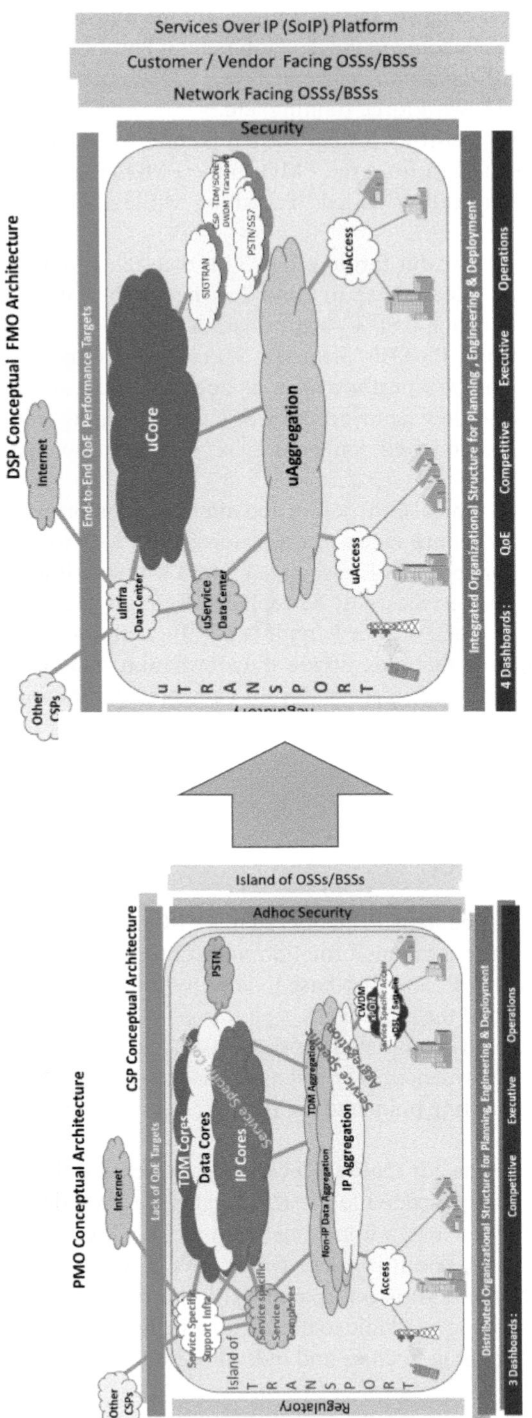

Figure 8.1. FNO conceptual architecture—gap closure.

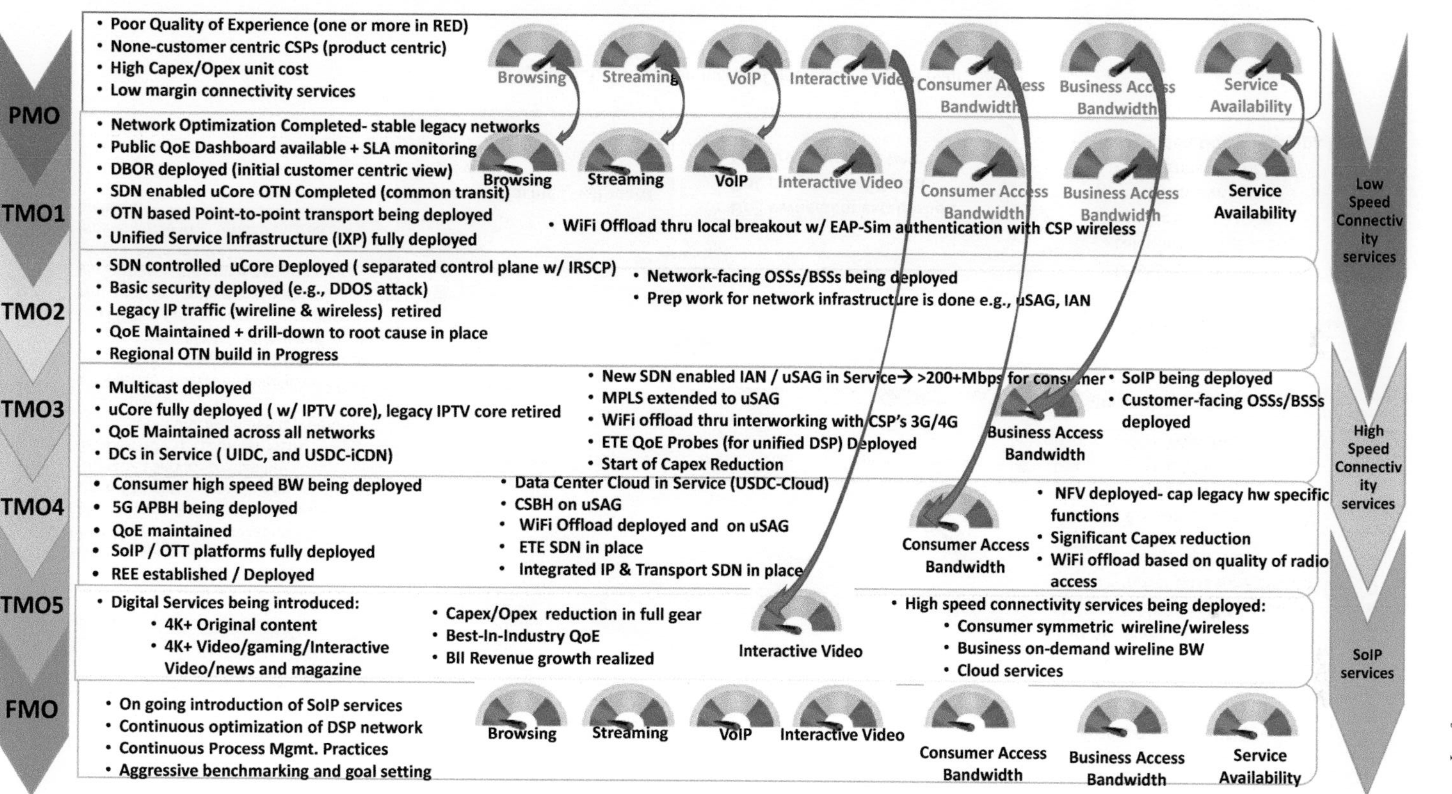

Figure 8.2. Technology tranfer from PMO to FMO.

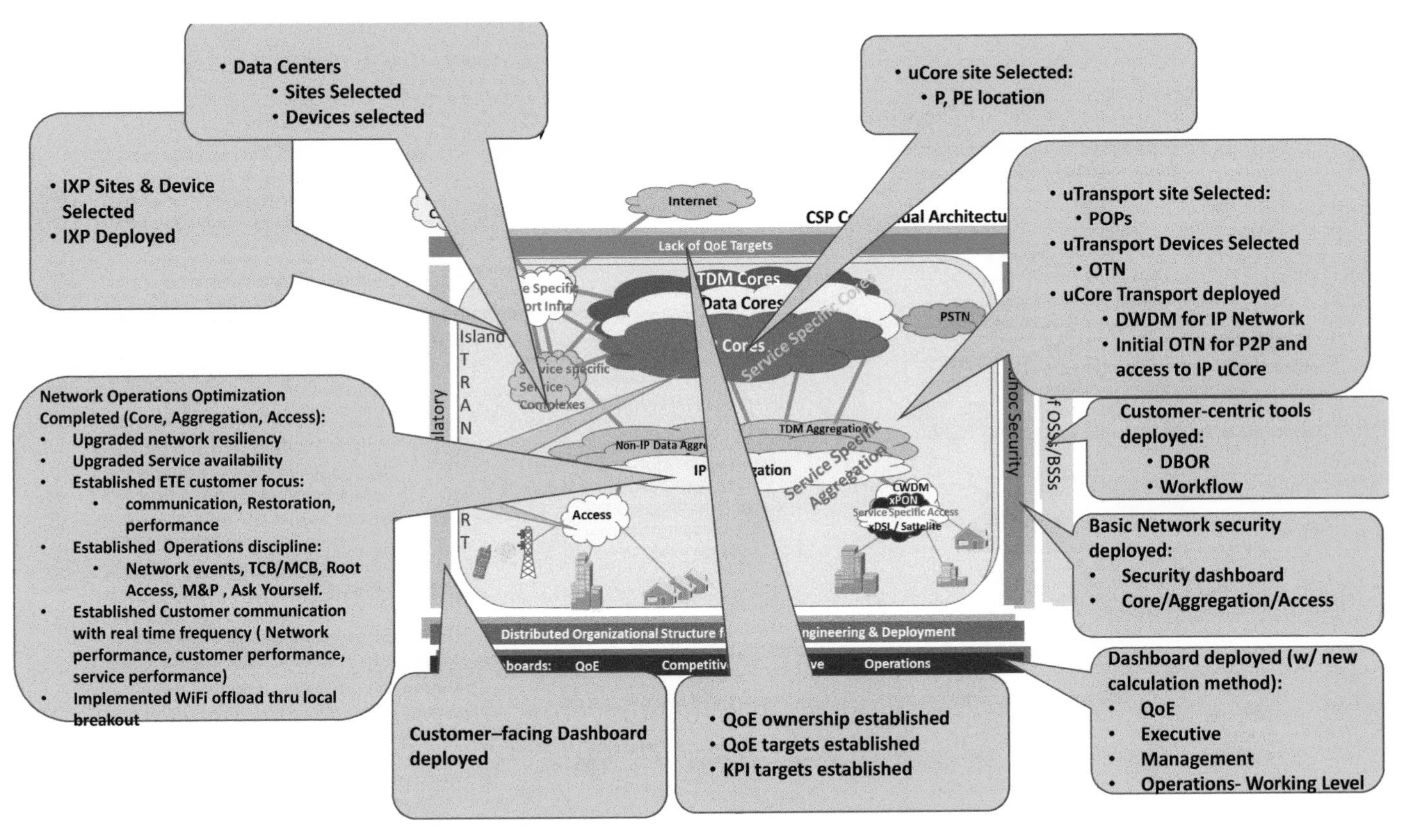

Figure 8.3. Transition mode of operation 1 action plan.

During PMO1, all legacy services are available and delivered with the best intention toward the QoE expectation,we are closing all performance gaps to the QoE and the design objectives by optimizing the legacy networks. This requires upgrading the network resiliency, upgrading service availability, and deploying operations discipline across all legacy networks. As we make progress with the gap closures, we must share the progress (e.g., the improvements in the QoE performance) with the customers in real-time. This customer communication is only effective if it is customized for the individual medium/large business customers, and generalized for small business and consumers.

After we bring the performance of the legacy networks in line with the QoE expectation during the peak hours, then it is time to start the journey for network transformation. In that, we attempt to identify the uCore sites for the "P" nodes, as well as the initial set of "PE" sites. Remember that the "P" nodes are permanent locations and should last for decades, and the "PE" nodes are expanding with the growth of the population over time. Next, we start selecting the transport nodes feeding the "P" nodes and the initial set of the "PE" nodes. Also, it is critical to begin to establish the mesh transport network node plan for P2P interconnection and the connectivity to the uCore.

At this stage, we must deploy DBOR and establish APIs interfaces from all legacy OSSs/BSSs to the DBOR to ensure a consistent view of customers, products, billing, etc.

Also, it is critical to establish a basic security framework across all legacy networks to manage DDOS attacks as well as security for the users and the content.

Next, we start deploying the unified transport network, in that, we deploy the DWDM network for the IP network, followed by the OTN nodes for P2P and access to IP uCore. The build-out of the transport OTN node must be staged and prioritized region by region, consistent with the plan for the network growth in each market.

At the end of TMO1, the CSP must have stabilized the legacy networks for the key service flows for browsing, streaming, and VoIP. The wireless connectivity is supported through WiFi offload through local breakout with EAP-SIM authentication between the WiFi network and the CSP's 3G/4G network. The service available for both connectivity services as well as DSP services must be deliverable at five 9's and/or SLA-based level.

The customers must be able to monitor the performance of their services in real-time. At the same time, the customer has a one-view of all DSP services, and must be receiving one bill with significant improvement in the accuracy of their bill.

Also, the uCore nodes locations are fully identified, the technology for uCore routing complexes as well as uCore transport is decided, lab tested, and approved for use. Also, uCore transport and the point-to-point mesh national transport are both deployed.

The picture below (Figure 8.4) outlines the state of the CSP network at the end of TMO1.

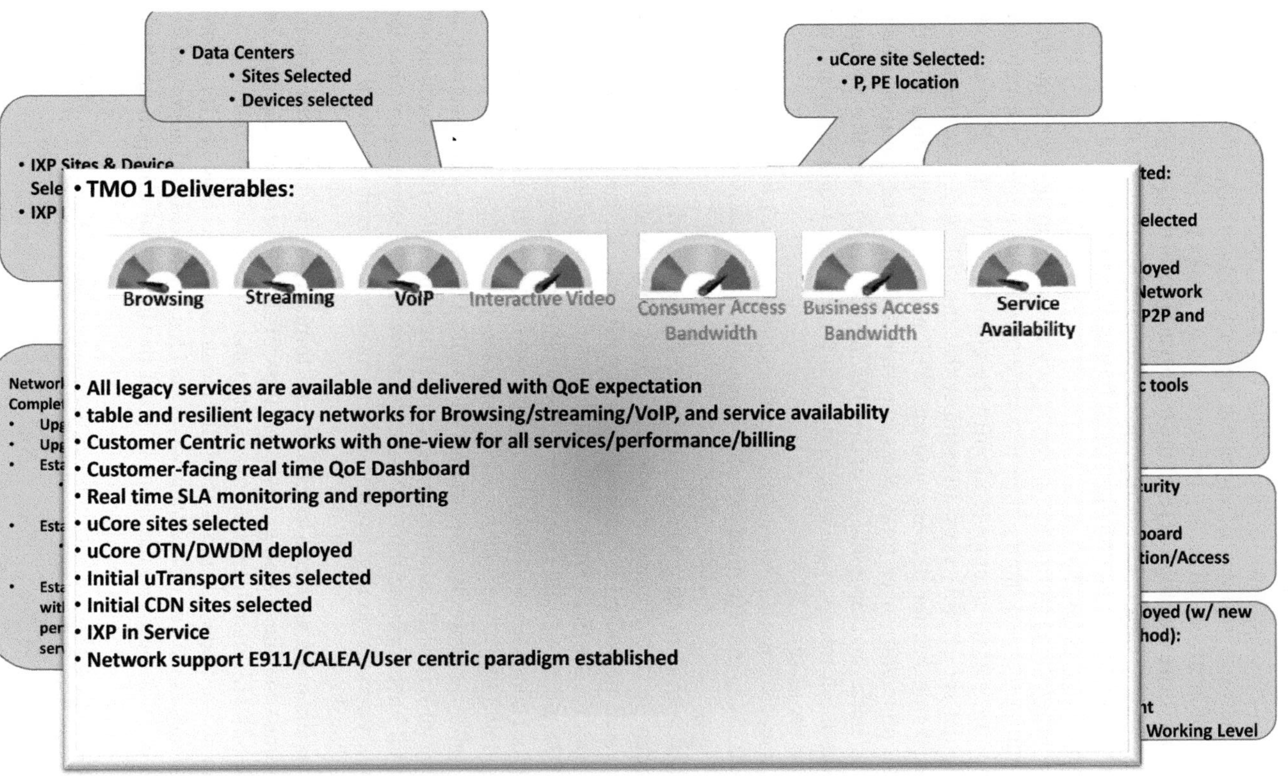

Figure 8.4. Transition mode of operation 1 action plan.

8.3 Transition Mode of Operation 2

TMO2 is the start of building the DSP infrastructure and platforms (Figure 8.5). The first and foremost task is to build the unified core network and to migrate wireless and wireline core backbones traffic into it.

In this phase, the prerequisite is that all legacy services are available and delivered with QoE expectation,we are now starting to build a unified national IP backbone network from scratch or leverage one of the existing core backbones to be the starting point for a unified core. This unified network (as outlined in the uCore section) must be built with significant size to handle the consolidation of all legacy core networks for the peak traffic, plus the capacity under failure mode, and the projected growth. This network could require the deployment of 20 to 35 “P” nodes in a 5000-kilometer national footprint, with 50–100+ core routers in all the national “P” nodes. It also requires building 100’s of 40/100+ Gbps transport routes (DWDM) to complete the partial mesh uCore with a load factor of ~ 1.5. In many cases, we may have to upgrade the fiber network routes to enable the capacity increase from low 10 Gig to 100+ Gbps.

At this step, the Unified Infrastructure Data Centers are deployed throughout the CSPs’ footprint at the National/Regional level.

The network architecture, no matter how complete it is planned for, must be prepared to change and adjust for latency, survivability and efficiency targets. Remember, all these changes must go on while we are trying to maintain the capped/legacy networks performing at the desired QoE performance level…this puts immense pressure on the legacy vendors to deliver circuit pack cards, chassis for the legacy networks while building the new one.

Once the uCore is partially completed and it is connected to the cooperating internet networks through the IXPs, it is time to start migrating the traffic from the legacy wireless networks, video network, as well as wireline networks to this unified backbone and to retire the legacy core backbones. This objective is achievable through the rehoming of the existing legacy networks’ PEs to the uCore. This task is extensive as the migration process could easily involve service by service migration, such as VPNs, MISs, wireless, etc., as the network latency expectations must be managed tightly to ensure adherence to the SLA as well as the QoE. In many cases the migration process will be gated, at times, by the build plan for layer 1 cross-sections, span protection, layer 3 node deployment and router protection, QoE dashboard implementation and/or real-time SLA compliance implementation. Another important gating factor is the preparedness of the OSSs/BSSs to manage the new uCore technologies.

Please keep in mind that the migration of the IPTV/broadcasting core is not advisable at this time because the ETE DSP network is not yet in place, nor is it enabled with the ETE Multicast capability. Any migration of such traffic has the potential to choke the ETE network.

While the uCore is being fully deployed, the uTransport network must be expanded to enable the connectivity of the PEs to the IP network, as well as a point-to-point mesh network. The uTransport network expansion must be prioritized based on the growth of the backhaul of the traffic to the PEs, as well as the demand for

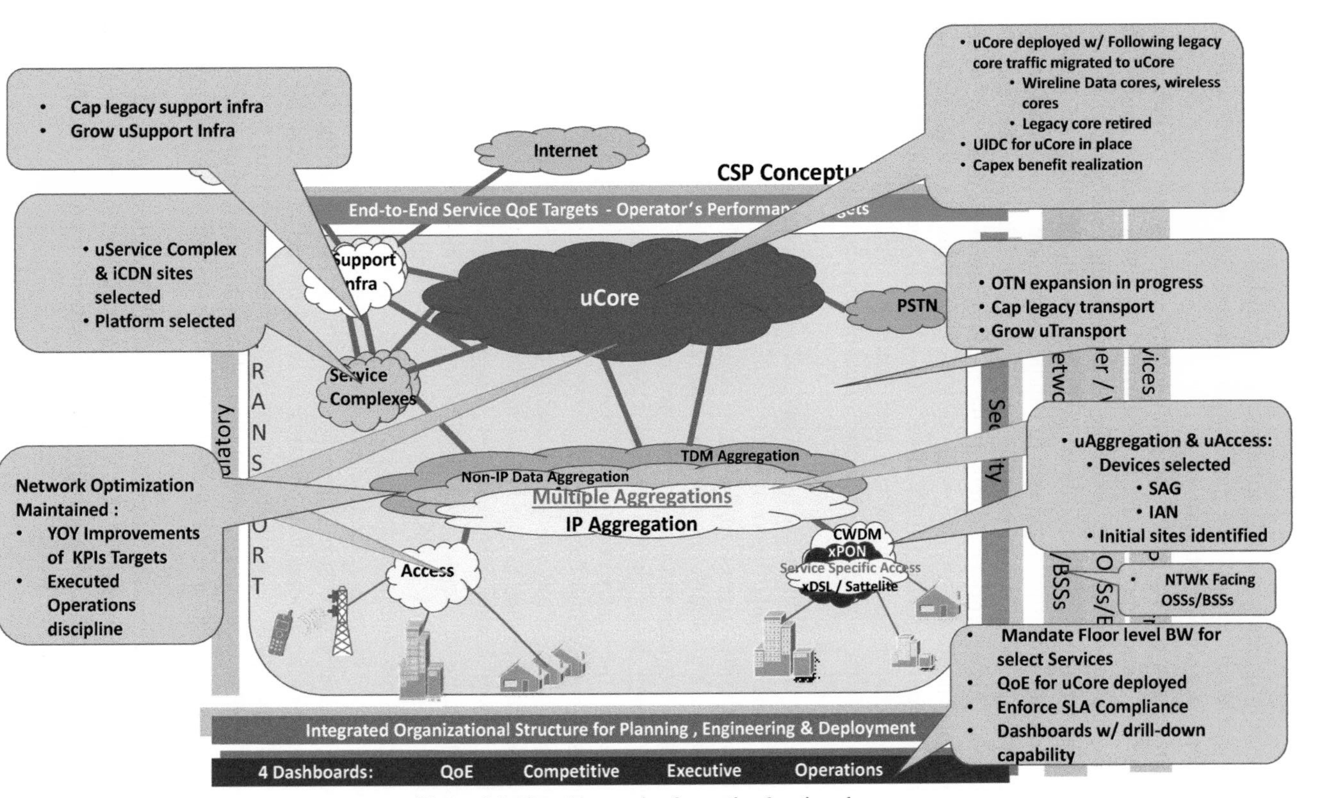

Figure 8.5. Transition mode of operation 2 action plan.

P2P connectivity at the national level. Any expansion of the uTransport must force capping of the expansion of the legacy transport in the same region/market to justify the expansion of the uTransport.

With the deployment of the uCore, the operations discipline must be expanded to cover the uCore. Needless to say that the legacy aggregations and the access must continue to be maintained through the operations discipline and network optimization techniques, as described before.

During the PMO2, care must be taken to select the technology for:

1. the uAccess (IAN) and uSAG devices, as well as the initial sites for the deployment of these technologies.
2. uService complex platforms (USDC) selected, as well as the initial iCDN sites.
3. uSupport infrastructure devices selected (UIDC), deployed, and the legacy nodes capped.

At the end of TMO2, the CSP must have deployed the uCore for wireless and wireline services with full retirement of the associated backbone cores. The QoE dashboard fully deployed for uCore with the drill-down capability into the legacy aggregation and access networks. Also, the regional OTN network is in full gear for expansion. The CSP has decided on the selection of the platforms for aggregation as well as the access networks.

Here is what to expect from the completion of the TMO2 in terms of performance and capabilities being deployed (Figure 8.6):

8.4 Transition Mode of Operation 3

TMO3 (Figure 8.7) is the key T-Plan step for the start of the unification of the DSP infrastructure.

In this step, All legacy services are available and delivered with QoE expectation, uCore is fully operational to include the unification of the broadcasting services over uCore, will begin the build-out and the unification of the aggregation and the access networks, will have full implementation of the Multicast capability throughout the ETE DSP unified network, which is the key capability to unify the IPTV/Broadcasting legacy network, will deliver the aggressive expansion of the uTransport (P2P and connectivity to the uCore) at the national level, consolidation of the legacy POP infrastructure nodes into the UIDC complexes, and deployment of the ETE dashboard for the unified ETE DSP network. Also, it is time to up the wireless connectivity for the last mile to WiFi offload through interworking with the CSP's 3G/4G/5G network.

TMO3 aggressively builds the mesh uTransport network at the national level. This network is the foundational block of capability to build the direct P2P services on it, as well as the backhaul to the IP network. This task is capital intensive and time-consuming, and will be completed over the TMO3 through TMO5. In this PMO3, we select a legacy ring-based metro network (e.g., SDH, Sonet), one by one, and convert it to a mesh OTN network over time. In that, replace select ADMs with OTN switches and start building the mesh transport without disrupting the existing

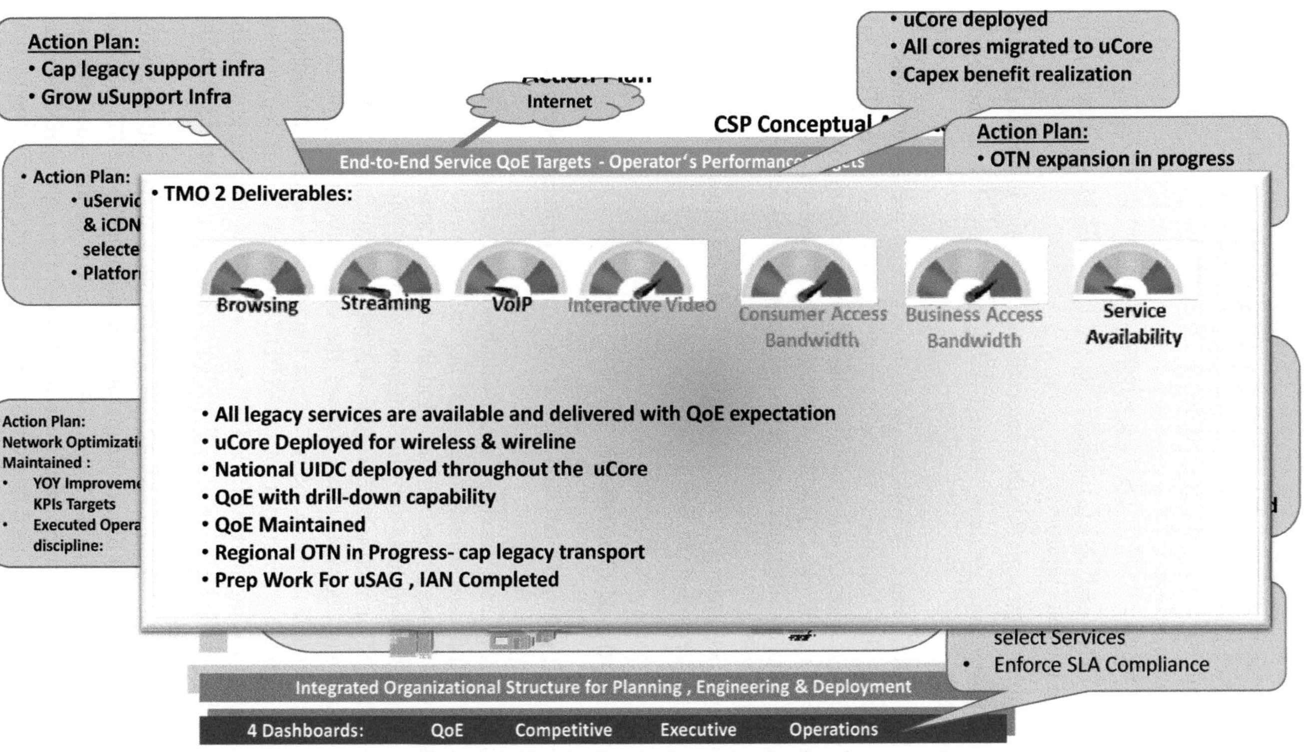

Figure 8.6. Transition mode of operation 2 action plan.

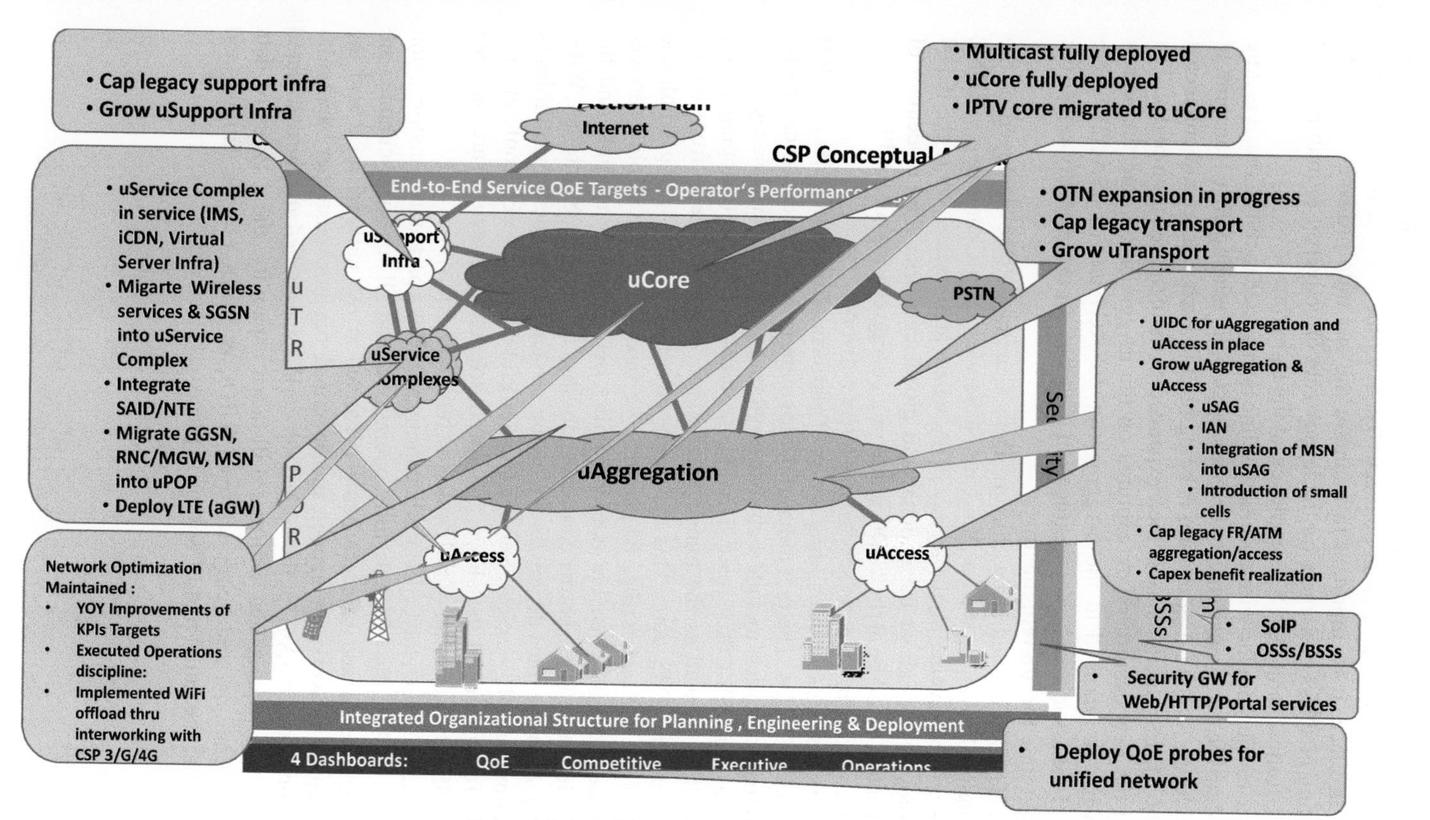

Figure 8.7. Transition mode of operation 3 action plan.

services. Over time, all ADMs will be replaced with the OTN switches and new OTN nodes will be added to ensure mesh connectivity. Also, the planned mesh transport will be put in place. It is important to maintain the last mile protection capacity for the existing business customers on the legacy ring or convert it to multiple P2P connections, from separate UIDC complexes, at the right time.

Once the traffic from the wireless and wireline is migrated to the uCore, it is time to start building the uAggregation network step by step, leveraging the uTransport, at the national level. As the uAggregation is being built in each market and connected to uCore, the MPLS functionality is then expanded to the uAggregation nodes. The growth and the new customers are captured on the uAggregation, and the legacy access network is rehomed to the uAggregation (uSAG1 and/or uSAG2) where possible, and the associated legacy aggregation is capped for any capacity expansion until the right time for full retirement.

During TMO3, Unified Infrastructure Data Centers will be deployed at the National/Regional level, as well as the Unified Service Data Center (e.g., iCDN). The UIDCs will be leveraging the legacy POPs where appropriate, while the iCDN will be deployed throughout the national and global networks. The placement of these data centers could be at the UIDC nodes (e.g., uAggregation) or the PEs, or the POPs at the cooperating internet service providers network.

At TMO3, one must start building the SoIP platform. This work starts at TMO3 and completes at TMO4. The first order of deployment is the unified session control and management that must include web core services, e.g., web service management, and SIP core, e.g., I/S-CSCF (call signaling protocols), ENUM (mapping phone number to URI), as well as the transit functions to directly route to/from border elements bypassing IMS core. This capability is critical for providing wholesale services for low featured traffic. In parallel, we must build and deploy service-specific border control to enable interoperability between the endpoints and the application elements. These border control capabilities include PSTN, internet, WiFi, and wireless. Also, we must deploy the following service management capabilities as a starting point for SoIP services: Service creation and delivery, voice, conferencing, messaging. To support all the above, the specific functionalities that are outlined under the OSSs/BSSs section of SoIP must be built and deployed.

At the end of TMO3, the CSP starts becoming a DSP as a part of its transformation (Figure 8.8). In that, the uCore is fully operational and all legacy backbones are retired. uAggregation and uAccess being expanded at the national level and the associated legacy networks are capped and/or retired, business customer bandwidth is available on-demand, QoE for the unified network is maintained and the associated dashboard is fully operational with the drill-down capability, the 2 types of data centers are operational and being deployed (UIDC & USDC) at the national and global level, and the initial part of SoIP is in place.

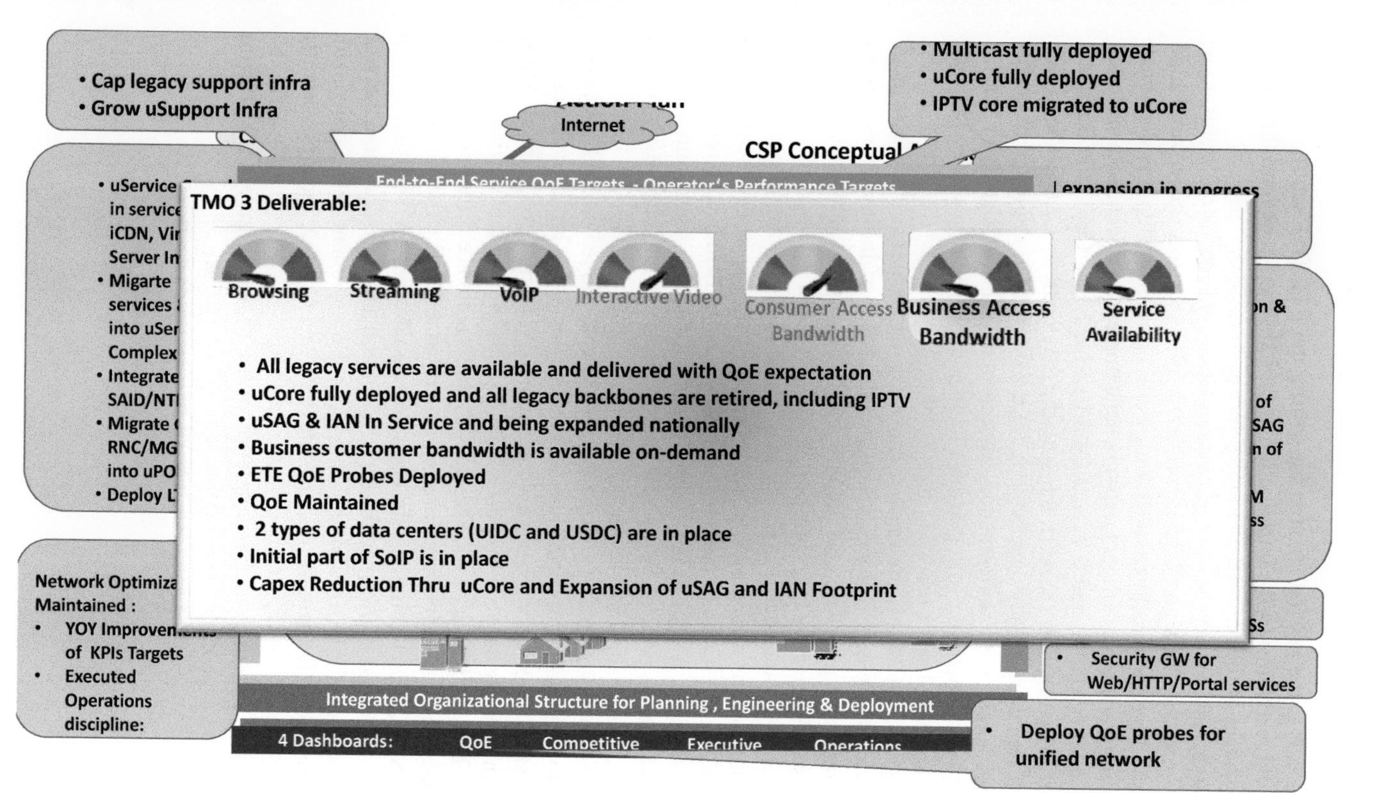

Figure 8.8. Transition mode of operation 3 action plan.

8.5 Transition Mode of Operation 4

TMO4 (Figure 8.9) is the key T-Plan step for getting the ETE DSP unified network fully in place and ready to start the introduction of digital services beyond the legacy connectivity services at the national and global levels.

In this step, all legacy services are available and delivered with QoE expectation,also, given the significant deployment of OTN, Consumer wired high speed bandwidth (symmetric, guaranteed with minimum 100+ Mbps) is deployed, WiFi offload is upgraded to intelligent network selection and services, with selection based on the quality of the radio access. WiFi offload and Cell Site Backhaul (CSBH) are on uAggregation (uSAG), the scalable data centers (USDC-Cloud) are fully deployed, replacing legacy data centers, ETE SDN is fully deployed to enable custom network slices with performance commitment, NFV is deployed for the majority of network functions, SoIP platform is fully deployed to include OTT, and other subscription-based services.

TMO4 aggressively builds the mesh uTransport network at the national level. In that, we divide the entire national footprint into 100's of individual consumer and business markets, prioritize and deploy mesh uTransport into each market, leveraging the existing legacy ring transport. The OTN mesh network must be built, not only to support P2P and access to IP, but also to enable densification requirements for the 5G+ wireless services. It is key to ensure that the fiber in the metro area is capable of handling 40+ Gbps to enable 100's of Mbps bandwidth requirements for symmetric IoT services. Also, it is key to prioritize markets with significant business concentration to ensure priority build-up of the uTransport into the business data centers, macro cell sites, and wireless access points.

At the end of TMO4 (Figure 8.10), the CSP has fully deployed the DSP necessary infrastructure to begin offering digital services in addition to the legacy connectivity services. In that, DSP offers high-speed consumer BW (symmetric, 100's+ Mbps guaranteed with bursting capabilities into Gbps), ETE SDN is now in service, QoE is maintained for ETE unified network, the cloud Data Centers are fully deployed, the majority of the network-specific functions are converted to NVF, CSBH is now fully deployed on uSAG, select cell sites are deployed on uAccess, WiFi Offload is on uSAG, and it is upgraded to leverage the quality of radio access in real-time, service complexes in service for SoIP and the Run-Time Environment (e.g., download, decode, play, rank for CDN) on the end-user devices, and CAPEX reduction is fully being realized.

Here is what to expect from the completion of the TMO4 (Figure 8.10) in terms of performance and capabilities being deployed:

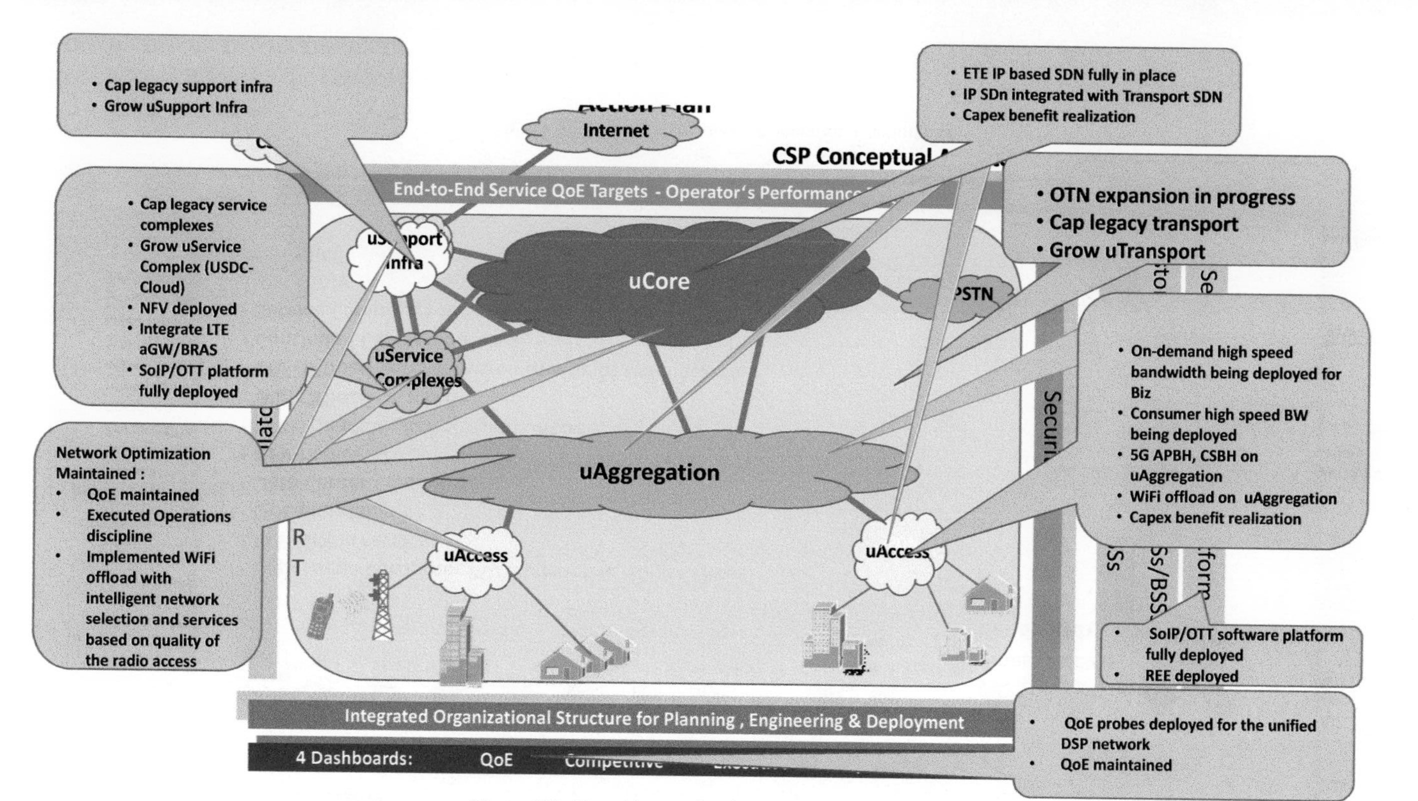

Figure 8.9. Transition mode of operation 4 action plan.

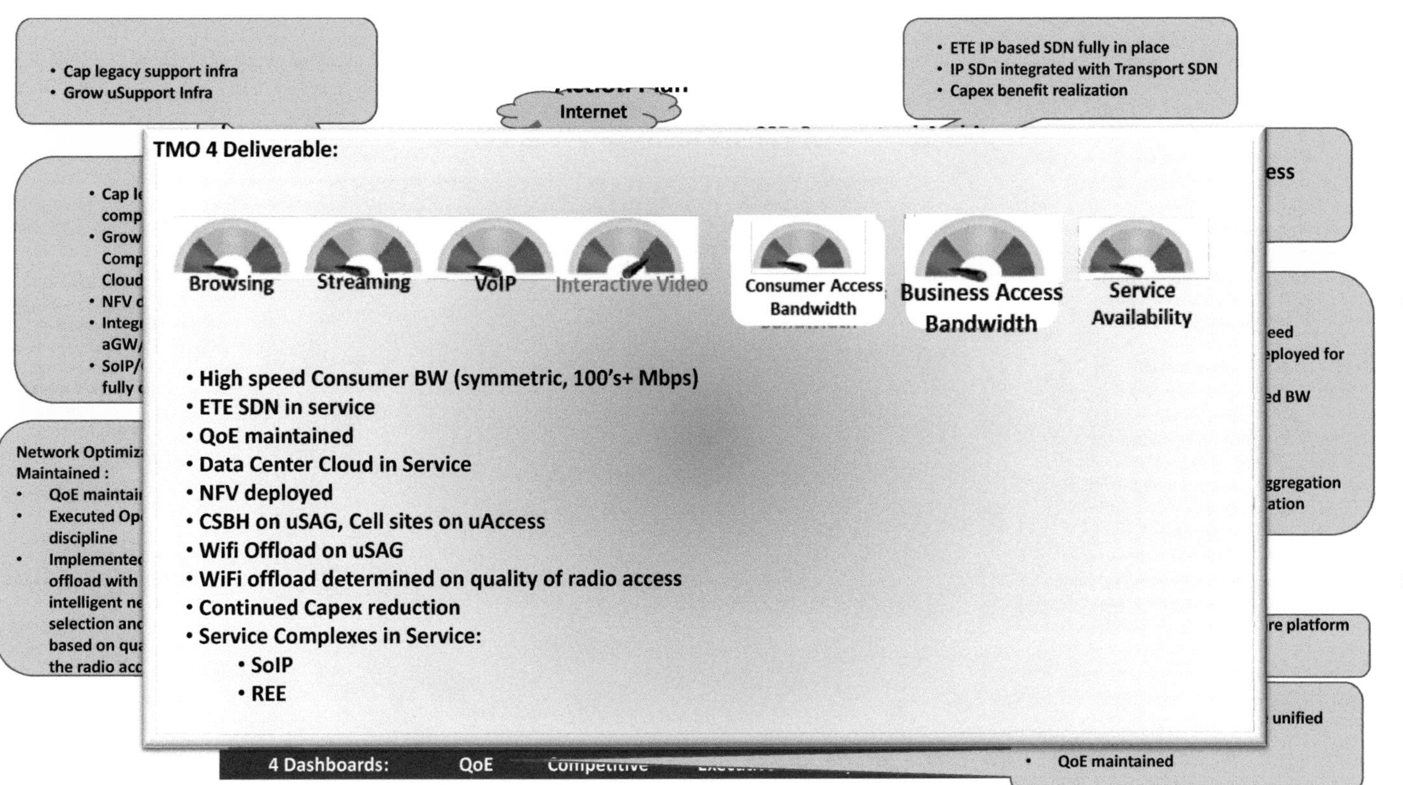

Figure 8.10. Transition mode of operation 4 action plan.

8.6 Transition Mode of Operation 5

TMO5 (Figure 8.11) is the final step in the transformation of a CSP into a DSP. In this step, DSPs achieve the full-scale objective of the transformation by the introduction of digital services, such as video subscription, interactive video such as gaming, news-and-magazine subscription services, and a continuous flow of original content at the national and the global scale.

In this step, all legacy services are capped and the growth and new customers utilize the new digital services on the DSP network. The agile process is in place for the rapid introduction of new services: high-speed connectivity and digital services.

TMO5 will continue to expand the OTN network. This network is the primary means for point-to-point transport services and access for IP and internet services. Also, the uAggregation and uAccess network will be the target for expansion. The unified data centers will continue to grow in space and numbers. For example, the iCDN data centers (USDC-iCDN) will grow in numbers as these data centers are being placed closer and closer to the eyeballs at the edge of the uAggegation nodes. Although the national UIDC is stable in terms of the number, the regional and distribution data centers (UIDC-regional/metro and distribution) will continue to grow in numbers to provide better and better geographical coverage. Also, the USDC-Cloud data centers will continue to expand in size and numbers in order to support global coverage.

At the end of TMO5 (Figure 8.12), the DSP enjoys best-in-class revenue growth, high margin new services, it provides BII QoE and SLA performance, high-speed connectivity services (wired and wireless), and the BII CAPEX/OPEX unit cost structure (Figure 8.12).

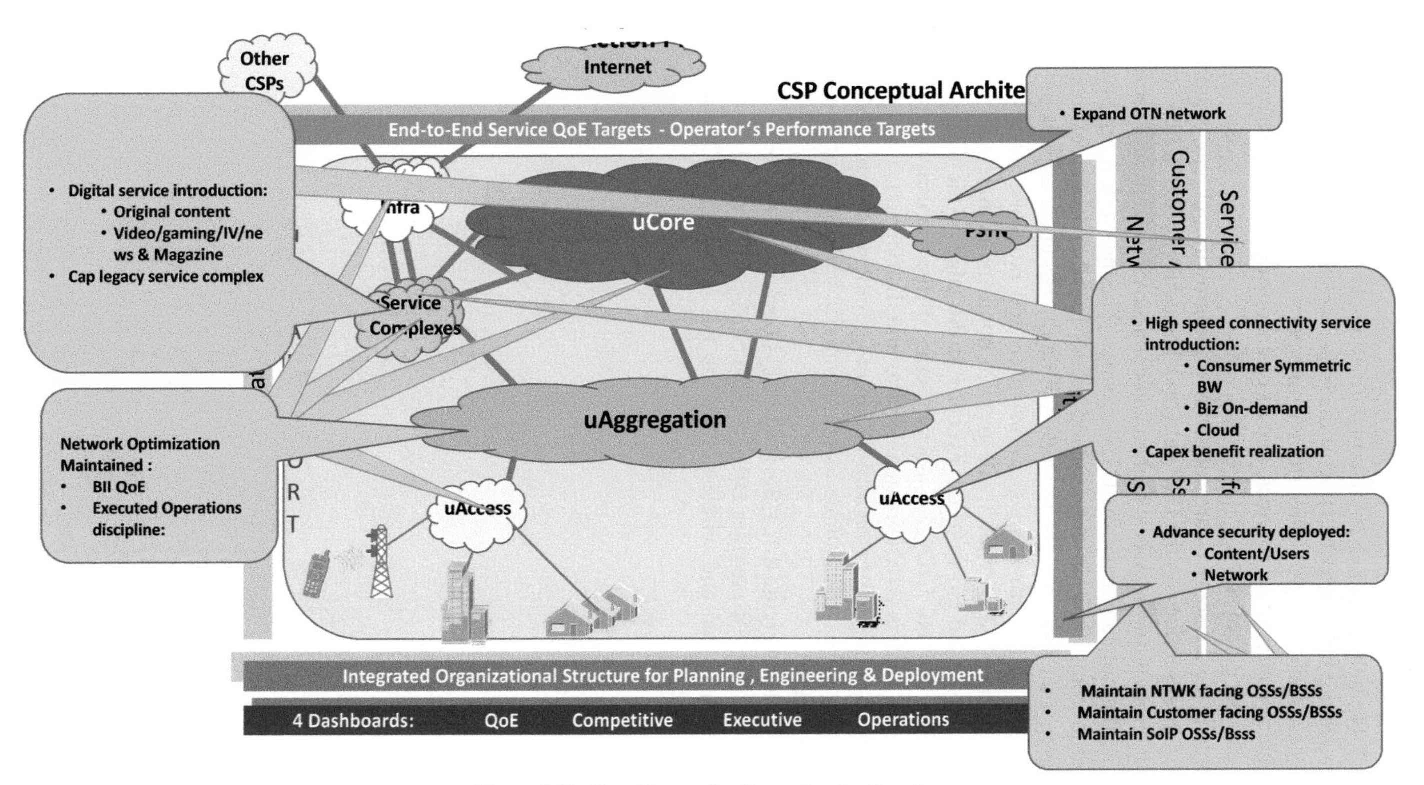

Figure 8.11. Transition mode of operation 5 action plan.

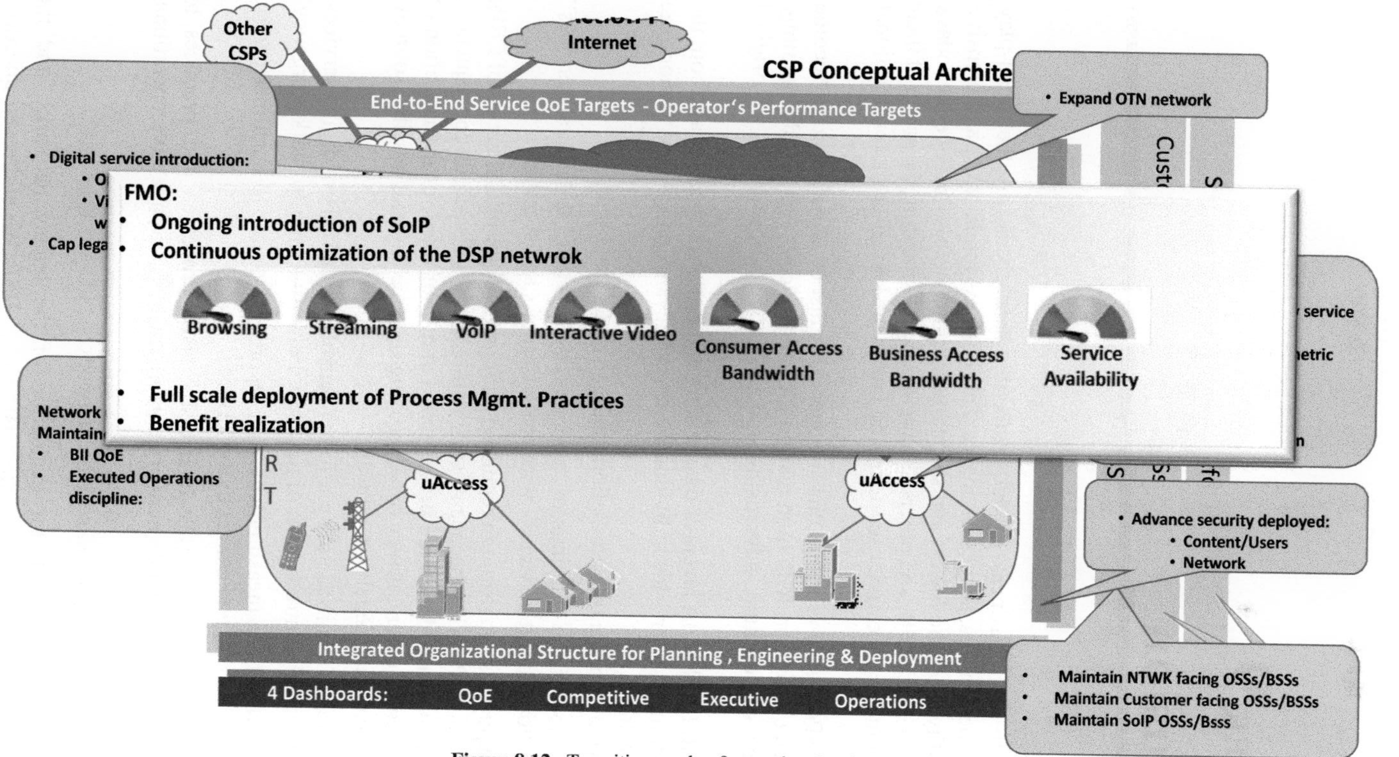

Figure 8.12. Transition mode of operation 5 action plan.

CHAPTER 9
Conclusion

The traditional CSP business model, with the sole focus on the quad-play connectivity-pipe services, is not sustainable. This business model is capital intensive and a commodity, with the lowest and shrinking margin. We also believe the todays companies like Google, FB, and others are aggressively expanding the scope of their offerings to include connectivity services, and that is the kiss of death for traditional CSPs.

For the CSPs to stay relevant, this arrangement must be transformed to become a Digital Service Provider, with high margin digital services such as video/gaming/news/magazine subscription, advertising and continuous introduction of original content.

A successful DSP must leverage data mining and analytics to know all about the users/end-customers, to provide individualized and relevant content, and to monetize the user engagement.

In that, we outlined a transformation roadmap and a playbook:

(1) Redefinition of the offerings of "connectivity services" to "digital services" and the "touchpoint services"- with end customers and suppliers.

(2) Content purification (through regulated arrangement) for liberalization of CSPs to become DSPs. The absence of such regulation has disastrous consequences as the unregulated DSPs are programmed and prioritized for monetization of end-user engagement, driven by advertising revenue from all possible sources: legal/ethical and illegal/non-ethical.

(3) Introduction of a unified SoIP platform to enable the delivery of digital services, and to enable the introduction of new services with a time-to-market urgency.

(4) Redefinition of the measurements and reporting systems to customer-centric QoE for all digital and connectivity services.

(5) Adaptation of timeless Best-In-Industry (BII) processes and practices to ensure an ETE sustainable network performance at a competitive operational efficiency.

(6) Transformation to a unified scalable network.

We provided a playbook for evaluation of the legacy networks, and the transformation of its shortcomings into a BII global DSP. We addressed questions about the "what" and "why" of transformation. The answers to the "how-to" of transformation are not the subject of this book as they are highly dependent on the state of the CSP before the start of the transformation. The "how-to" is best learned and practiced on the job, motivated by the need for transformation, and leveraging the partnership with experts who have global transformation experience.

Finally, we would like to emphasize that once your transformation is complete, your end-customers must not experience any more improvement in the network and service performance, but must experience competitively priced new services, significant scalability, a much broader base for coverage, and much faster time-to-market for new high margin services.

CHAPTER 10
Appendix

Acronym	Description
3CP	Command, Control, Communication
AAA	Authentication, Authorization and Accounting
ACL	Access Control List
ADM	Add-Drop Multiplexer
ANDSF	Access Network Discovery and Selection Function
API	Application Programming Interface
ARPU	Average Revenue Per User
ASY	Ask Your Self
BAU	Business As Usual
BGP	Border Gateway Protocol
BII	Best-In-Industry
BIZ	Business
BRAS	Broadband Remote Access Server
BSS	Business Support System
BU	Business Unit
CA	Certificate Authority
CALEA	Communications Assistance for Law Enforcement Act
Capex	Capital Expense
CPE	Customer Premise Equipment
CPPU	Customer Performance Processing Units
CDN	Content Distribution Network
CDR	Call Detail Record
CIO	Chief Information Officer
CFO	Chief Operating Officer
CO	Central Office
Co0	Concept of Zero

Co1	Concept of One
CoN	Concept of None
CoS	Class of Service
CSBH	Cell Site Backhaul
CSCF	Call Session Control Function
CPU	Central Processing Unit
CTO	Chief Technology Officer
CSP	Communication Service Provider
CWDM	Coarce Wavelength Division Multiplexing
DaaS	Data as a Service
DBOR	Data Base of Record
DDOS	Distributed Denial of Service
DL	Down Link
DMOQ	Direct Measure of Quality
DNS	Domain Name System
DO	Duty Officer
DPI	Deep Packet Inspection
DPM	Defects Per Million
DSP	Digital Service Provider
DWDM	Dense Wavelength Division Multiplexing
E911	Enhanced 911
EAG	External Application Gateway
EAP	Extensible Authentication Protocol
EBITDA	Earning Before Income Tax and Amortization
EF&I	Engineered, Furnished, and Installed
E-LAN	Ethernet Local Area Network
ENUM	E.164 Number to URI Mapping
EOCU	Ethernet over Copper
FHRP	First Hop Redundancy Protocol
ETE	End To End
FMO	Future Mode of Operation
fps	Frame Per Second
FPS	First Person Shooter
FRR	Fast Reroute
FTTN	Fiber To The Node
GGSN	GPRS Support Node
GW	Gate Way
HD	High Definition

HLR	Home location register
HSIA	High Speed Internet Access
HTTP	HyperText Transfer Protocol
HW	Hardware
IAN	Integrated Access Node
IaaS	Infrastructure as a Service
iCDN	Intelligent Content Distribution Network
ICMP	Internet Control Message Protocol
IDS	Intrusion Detection System
I-CSCF	Interogating CSCF
IGP	Interior Gateway Protocol
IMS	IP Multimedia Subsystem
IoT	Internet of Things
IP	Internet Protocol
IP	Intellectual Property
IPA	Intelligent Personal Assistant
IPDSLM	Internet Protocol Digital Subscriber Line Access Multiplexer
IRSCP	Intelligent Routing and Service Control Platform
IV	Interactive Video
IXP	Inter Exchange Provider
KPI	Key Performance Indicator
LI	Lawful Intercept
LTE	Long Term Evolution
M&P	Methods and Procedures
MAG	Media Application Gateway
MCB	Master Control Bridge
MIS	Managed Internet Service
MSC	Mobile Switching Center
MME	Mobility Management Entity
MSA	Multi Service Access
MSF	Macro Service Flow
MTTR	Mean Time To Repair
NEP	Network Equipment Provider
NFV	Network Functions Virtualization
NOC	Network Operations Center
NPL	Network Performance Level
Opex	Operating Expense
Ops	Operations

OSP	Out Side Plant
OSPF	Open Shortest Path First
OSS	Operations Support Systems
OT	Outage Management
OTT	Over The Top
P	P Router
P2MP	point to multi-point
P2P	point to point
PaaS	Platform as a Service
PCRF	Policy and Charging Rules Function
PE	Provider Edge Router
PKI	Private Key Infrastructure
PMO	Present Mode of Operation
POE	Plant Operating Error
PON	Passive Optical Network
POP	Ponit of Present
POR	Plan Of Record
POTS	Plain Old Telephone Service
PSTN	Public Switched Telephone Network
QoE	Quality of Experience
QoS	Quality of Service
Qplan	Quarterly Plan
RAN	Radio Access Network
REE	Runtime Execution Environment
RNC	Radio Network Controller
ROI	Return On Investment
ROIC	Return on Invested Capital
ROW	Right Of Way
RR	Route Reflector
RSVP	Resource Reservation Protocol
RTT	Round Trip Time
SC&D	SoIP Service Creation and Delivery
S-CSCF	Serving CSCF
SD	Simple Definition
SDH	Synchronous Digital Hierarchy
SDK	Software Development Kit
SDN	Software Defined Network
SDR	Service Disaster Recovery

SGSN	Serving GPRS Support Node
SHO	Super Hub Office
SIAD	Service Integrated Access Device
SIM	Subscriber Identification Module
SIGTRAN	Signaling Transport
SLA	Service Level Agreement
SLT	Senior Leadership Team
SME	Subject Matter Expert
SMOP	Service Method of Procedure
SoIP	Service over Internet Protocol
SS7	Signaling System No. 7
SSL	Secure Sockets Layer
SW	Software
TC	Transformation Culture
TCB	Technical Control Bridge
TDM	Time Division Multiplexing
TE	Traffic Engineering
TLS	Transport Layer Security
TMO	Transition Mode of Operation
Tplan	Time Plan
TTR	Time To Repair
uAccess	Unified Access
uAggregation	Unified Aggregation
uCore	Unified Core
UDP	User Data Protocol
UL	Up Link
uPOP	Unified POP
URI	Uniform Resource Identifier
uSAG	Unified Service Aggregation
USB	Universal Service Bus
USDC	Unified Service Data Center
uService	Unified Service
UIDC	Unified Infrastructure Data Center
uTransport	Unified Transport
VHO	Video Hub Office
VOD	Video On Demand
VPLS	Virtual Private LAN Service
VPN	Virtual Private Network

VS	Vital Sign
WCSMP	WiFi Core Service Management Platform
xDSL	x Digital Subscriber Line (of any type)
xPON	x Passive Optical Network (of any type)
WR	War Room
YOY	Year over Year

CHAPTER 11
References

A future for mobile operators: The keys to successful reinvention. (2017, Feb 8). Retrieved from McKinsey &Company: a-future-for-mobile-operators-the-keys-to-successful-reinvention (2).pdf.

Adtran. (2020). IP Multi-service Access Node. Retrieved from adtran.com: https://portal.adtran.com/pub/Library/Data_Sheets/International_/I61132006G1-8_hiX5630.pdf.

Akamai. (2020). Content Delivery for an Evolving Internet: Choosing the Right CDN for Today & Tomorrow. Akamai.

Almes, G. E. (1998). A One-Way Delay Metric for IPPM. Internet Engineering Task Force.

Aruba. (2020). INTEGRATING WI-FI AND. Sunnyvale: www.arubanetworks.com.

at&t. (1999). Communication, Command, and Control. at&t.

at&t. (2000). Ask YouSelf. at&t.

at&t. (2003). Concept-of-None. at&t.

at&t. (2003). Concept-of-One. at&t.

at&t. (2003). Concept-of-Zero. at&t.

at&t. (2003). Data Base of Record. at&t.

at&t. (2010). AT&T Common Architecture for Real Time Services (CARTS). at&t.

Balakrishnan, H. (2009). Wide-Area Internet Routing. Massachusetts Institute of Technology-Department of Electrical Engineering and Computer Science.

Barman, N. and Martini, M. G. (2019, March 13). QoE Modeling for HTTP Adaptive Video Streaming–A Survey and Open Challenges. IEEE Access, pp. 30831–30859.

Begen, A. C. (2012). IPTV, Internet Video and Adaptive Streaming Technologies. Cisco Canada Plus.

Benjamin, S. M. and Speta, J. B. (2019). Internet and Telecommunication Regulation. Carolina Academic Press.

Bitmovin. (2018). Digital Rights Management. bitmovin.com.

Bouraqia, K. E. (2019). Quality of Experience for Streaming Services: Measurements, Challenges and Insights. IEEE.

Breuer, D. E. (2013). Unified Access and Aggregation Network Allowing Fixed and Mobile Networks to Converge. Optical Fiber Communication Conference/National Fiber Optic Engineers Conference. OSA Technical Digest (online).

Butler, B. (2016, 2 10). SD-WAN: What it is and why you'll use it one day. Retrieved from https://www.cio.com/article/3032334/sd-wan-what-it-is-and-why-youll-use-it-one-day.html.

Capacity and Performance Management: Best Practices White Paper. (2005, October 4). Retrieved from Cisco: https://www.cisco.com/c/en/us/support/docs/availability/high-availability/20769-performwp.html.

Chen, J. E. (2012). Customer experience oriented service quality management. IEEE Symposium on Robotics and Applications (ISRA).

Choudrie and Middleton. (2013). Management of Broadband Technology and Innovation. books.google.com.

Christensen, C. (2016). Innovator's Dilemma. Harvard Business Review Press.

Chu, Y.-H. E. (2014). Software-defined QoE measurement architecture. IEEE.

CISA. (2019, November 20). Understanding Denial-of-Service Attacks. Retrieved from Cybersecurity & Infrastructure Security Agency: https://us-cert.cisa.gov/ncas/tips/ST04-015#:~:text=What%20is%20a%20denial%2Dof,a%20malicious%20cyber%20threat%20actor.

Cisco. (2005). Service Level Management: Best Practices White Paper. Cisco.

Cisco. (2009). Advanced Topics in MPLS-TE Deployment. Cisco IOS and NX-OS Software.

Cisco. (2016). Converged Transport Architecture: Improving Scale and Efficiency in Service Provider Backbone Networks. Cisco Carrier Routing Systems.

Cisco. (2020). MPLS Traffic Engineering Fast Reroute—Link Protection. Retrieved from Cisco IOS Software Releases 12.0 ST: https://www.cisco.com/en/US/docs/ios/12_0st/12_0st10/feature/guide/fastrout.html#wp1015327.

Clinckx, Nicolas and Baffalio, Yann. (2014). Optimize network Opex and Capex while enhancing the quality of service. Telecom, Media and Technology.

Costantini, P. (2011). Cash Return on Capital Invested. Elsevier Science.

Dai, D. (2011). A Survey of Quality of Experience. Meeting of the European Network of Universities and Companies in Information and Communication Engineering (pp. 146–156). Berlin: Springer.

Demichelis, C. (1998). Instantaneous Packet Delay Variation Metric for IPPM. Internet Engineering Task Force.

encoding.com. (2020). The complete guide to Apple Fairplay. Retrieved from encoding.com: https://www.encoding.com/apple-fairplay/.

Flavel, A. E. (2015). FastRoute: A Scalable Load-Aware Anycast Routing Architecture for Modern CDNs. USENIX Symposium on Networked Systems Design and Implementation. Oakland: USENIX Symposium.

Frank, B. E. (2013, July). Pushing CDN-ISP Collaboration to the Limit. ACM SIGCOMM Computer Communication Review.

Fujitsu. (2010). Tthe Key Benefits of OTN Networks. Retrieved from www.fujitsu.com: OTNNetworkBenefitswp.pdf.

Google. (2005). Understanding Changing Telecommunications: Building a successful Telecom Business. Books.google.com.

Goransson, P., Black, C. and Culver, T. (2016). Software Defined Networks (a comprehensive approach). Elsevier Science.

Greene, W. and Lancaster, B. (2006). Carrier-Grade: Five Nines, the Myth and the Reality. Pipeline, pp. Volume 3, issue 11.

Halper, H. (2010). A Guide to Building a Metrics-Driven Organization. Needham: Hurwitz & Associates.

Harmon, P. (2002). Business process change. Elsevier.

Hopkinson, M. (2017). Net Present Value and Risk Modelling for Projects. Routledge.

Hossfeld, T. (2012). Waiting times in quality of experience for web based services. Fourth International Workshop on Quality of Multimedia Experience (QoMEX). Melbourne: IEEE.

Huber, M. and O'Gorman, Susanne. (2008). From Customer Retention to a Holistic Stakeholder Management System. Springer-Verlag Berlin Heidelberg. (2010, July). IBM Business Analytics for telcos. Ottawa: IBM Canada.

Institute Project Management. (2018). Benefits Realization Management: A Practice Guide. Kindle Edition.

Introduction to eTOM White Paper. (2017, Feb. 16). Retrieved from https://www.cisco.com/c/en/us/products/index.html: www.cisco.com.

Jacobs, Andy. (2020). The Steady Decline in Operating Cash Flow for Telcos. Retrieved from Subex: https://www.subex.com/the-steady-decline-in-operating-cash-flow-for-telcos/.

Jelassi, S. (2012, January 11). Quality of Experience of VoIP Service: A Survey of Assessment Approaches and Open Issues. IEEE Communications Surveys & Tutorials, pp. 491–513.

Josyula, Venkata et al. (2012). Cloud Computing: Automating the Virtualized Data Center. Indianapolis: Cisco Press.

Journals.elsevier.com/telecommunications-policy. (2020). OTT and live streaming services: Past, Present and Future. The International Journal of Digital Economy, Data Sciences and New Media.

Juniper. (2015). Building a Universal Access Network . Juniper.

Kahn, R. (1974). A Protocol for Packet Network Intercommunication. IEEE Transactions on Communications. IEEE Transactions on Communications.

Kenton, W. (2019). Discounted Payback Period Definition. Investopedia.

Kirstädter, A. E. (2006). Carrier-grade Ethernet for packet core networks. Asia-Pacific Optical Communications. Gwangju, South Korea.

Laghari, K. U. and Connelly, K. (2012). Toward total quality of experience: A QoE model in a communication ecosystem. IEEE, 58–65.

Lange, C. E. (2010). Effects of network node consolidation in optical access and aggregation networks on costs and power consumption. Society of Photo-Optical Instrumentation Engineers (SPIE). San Francisco, California, United States: SPIE OPTO.

Lema, M. A. (2017). Business Case and Technology Analysis for 5G Low Latency Applications. IEEE Access (Volume 5), 5917–5935.

Marr, B. (2012). Key Performance Indicators: The 75+ Measures Every Manager Needs to Know. Key Performance Indicators: The 75+ Measures Every Manager Needs to Know.

Mastel, M. S. (2003). A Complete Cost-Reduction Strategy for Your Corporate Telecommunications Bills. McGraw-Hill Professional; 1 edition.

Meakin, Rolf . (2012). Telecom capital expenditure (Capex). PwC.

Mellouk, A. and Mushtaq, M.-S. (2017). Quality of Experience Paradigm in Multimedia Services: Application to OTT Video Streaming and VoIP Services (Networks and Telecommunications. Elsevier.

NASPO Benchmarking Workgroup. (2007). Benchmarking Cost Savings & Cost Avoidance. National Association of State Procurement Officials.

Open IPTV Forum. (2013). OIPF Release 2 Specification Functional Architecture. Val; bonne: http://www.oipf.tv.

Orimolad, J. E. (2016). ANDSF-based WLAN Offloading in the EvolvedPacket System (EPS). 18th Mediterranean Electrotechnical Conference. Limassol: https://www.researchgate.net/publication.

Ovstoll, J. (2019). The Big Zero : The Transformation of ZBB into a Force for Growth, Innovation and Competitive Ad Vantage. Penguin Books, Limited.

Patterson, D. and Fox, A. (2014). Engineering Software as a Service: An Agile Approach Using Cloud Computing. Strawberry Canyon LLC.

RootMetrics. (2020). Retrieved from Connected insights for your connected lifestyle: https://www.rootmetrics.com/en-GB/home.

ROSSI, Luisa. (2014). Proposal for the reform of the regulation of digital services. White Paper- Proposal by the author.

Rossman, J. (2016). The Amazon Way on IoT. John E. Rossman.

Sabella, D., Reznik, A. and Frazao, R. (2019). Multi-Access Edge Computing in Action. books.google.com.

Serpanos, D. and Wolf, M. (2017). Internet-of-Things (IoT) Systems: Architectures, Algorithms. books.google.com.

Sheen, R. and Gall, A. (2015). HBR Guide to Building Your Business Case. Harvard Business School Publishing Corporation.

Sivalingam, K. M. and Subramaniam, S. (2006). Optical WDM Networks. Springer Science & Business Media.

Sowry, D. E. (2018). Edge Data Centers. elecommunications Industry Association's (TIA).

Stalk, J. G. and Hout, T. M. (1990). Competing Against Time : How Time-based Competition is Reshaping Global Markets. Free Press.

Sujata, J. e. (2015). Impact of Over the Top (OTT) Services on Telecom Service Providers. Indian Journal of Science and Technology, 8(S4): 145–160.

Taga, Karim et al. (2020). Cost Reduction in the Telecom Industry . Retrieved from Arthur D. Little: https://www.adlittle.com/sites/default/files/viewpoints/ADL_Cost_Reduction_Telecom_Industry.pdf.

Telecom Capital Expenditure. (2020). Retrieved from pwc Global: https://www.pwc.com/gx/en/industries/tmt/telecommunications/capex.html.

Tennant, G. (2001). SIX SIGMA: SPC and TQM in Manufacturing and Services. Gower Publishing, Ltd.

Tikhonov, V. I. (2013). Integrated telecommunication technology for the next generation networks. 2013 Proceedings of ITU Kaleidoscope: Building Sustainable Communities. Kyoto, Japan: IEEE.

Vincenzo Basile, Francesco Cotrone, Christoph Uferer and Maximilian Schaue. (2018). Monetizing investments in telecom network infrastructure. Arthur D Little.

Wieder et al. (2011). Service Level Agreements for Cloud Computing. Springer Science & Business Media.

Wrobel, L. and Wrobel, S. (2009). Disaster Recovery Planning for Communications and Critical Infrastructure (Artech House Telecommunications). Artech House.

Zakas, N. C. (2011, November 29). How content delivery networks (CDNs) work. Retrieved from Human Who Codes: https://humanwhocodes.com/blog/2011/11/29/how-content-delivery-networks-cdns-work/.

Index

DWDM 6, 49, 50, 79, 83, 114, 117, 120–124, 127, 129, 135, 182, 183, 220–223

E

F

G

H

I

J

L

M

T

U

V

W

Y

Z